T0201203

Composite Structures: Effects of Defects

Composite Structures: Effects of Defects

Rani Elhajjar
University of Wisconsin
Milwaukee, Wisconsin
USA

Peter Grant
Independent Aviation & Aerospace Professional
Medford, Oregon
USA

Cindy Ashforth
Federal Aviation Administration
Seattle, Washington
USA

The right of Rani Elhajjar, Peter Grant and Cindy Ashforth to be identified as the authors of this work has been asserted in accordance with law.

Registered Offices
John Wiley & Sons, Inc., 111 River Street, Hoboken, NJ 07030, USA
John Wiley & Sons Ltd, The Atrium, Southern Gate, Chichester, West Sussex, PO19 8SQ, UK

Editorial Office
The Atrium, Southern Gate, Chichester, West Sussex, PO19 8SQ, UK

For details of our global editorial offices, customer services, and more information about Wiley products visit us at www.wiley.com.

Wiley also publishes its books in a variety of electronic formats and by print-on-demand. Some content that appears in standard print versions of this book may not be available in other formats.

Limit of Liability/Disclaimer of Warranty
While the publisher and authors have used their best efforts in preparing this work, they make no representations or warranties with respect to the accuracy or completeness of the contents of this work and specifically disclaim all warranties, including without limitation any implied warranties of merchantability or fitness for a particular purpose. No warranty may be created or extended by sales representatives, written sales materials or promotional statements for this work. The fact that an organization, website, or product is referred to in this work as a citation and/or potential source of further information does not mean that the publisher and authors endorse the information or services the organization, website, or product may provide or recommendations it may make. This work is sold with the understanding that the publisher is not engaged in rendering professional services. The advice and strategies contained herein may not be suitable for your situation. You should consult with a specialist where appropriate. Further, readers should be aware that websites listed in this work may have changed or disappeared between when this work was written and when it is read. Neither the publisher nor authors shall be liable for any loss of profit or any other commercial damages, including but not limited to special, incidental, consequential, or other damages.

Library of Congress Cataloging-in-Publication Data

Names: Elhajjar, Rani, author. | Grant, Peter, 1942- author. | Ashforth,
     Cindy, author.
Title: Composite structures : effects of
     defects / Rani Elhajjar, University of Wisconsin, Milwaukee, Wisconsin,
     Peter Grant, Independent Aviation & Aerospace Professional, Medford,
     Oregon, Cindy Ashforth, Federal Aviation Administration, Seattle,
     Washington.
Description: First edition. | Hoboken, NJ : John Wiley & Sons, Ltd, [2019] |
     Includes bibliographical references and index. |
Identifiers: LCCN 2018026808 (print) | LCCN 2018030714 (ebook) | ISBN
     9781118997734 (Adobe PDF) | ISBN 9781118997727 (ePub) | ISBN 9781118997703
     (hardcover)
Subjects: LCSH: Composite construction. | Composite materials–Testing. |
     Product design.
Classification: LCC TA664 (ebook) | LCC TA664 .E44 2018 (print) | DDC
     624.1/8–dc23
LC record available at https://lccn.loc.gov/2018026808

Cover Design: Wiley
Cover Image: © blackjack/Shutterstock

Set in 10/12pt WarnockPro by SPi Global, Chennai, India

Printed and bound by CPI Group (UK) Ltd, Croydon, CR0 4YY

10 9 8 7 6 5 4 3 2 1

Contents

Preface

This book, unlike most other books on the subject of design of composites, deals directly with the practical implications of what can and often does go wrong during the fabrication and service of polymer-based composite structures. This book is written from the viewpoint of providing the practicing engineer a resource to design, build and certify a composite structure. Designing efficient composite structures requires a multidisciplinary awareness of complex chemical, physical and structural behaviors. Further, the idealizations of composite laminate design so often presented in classical lamination plate theory or failure theory discussions quickly run into problems. The assumptions based on these theories are often not capable of handling the realities of modern production processes and service environments.

While fiber-reinforced composites can offer superior properties when compared with other structural materials such as aluminum and steel, these property advantages can quickly diminish if composites are not carefully designed to account for the practical implications of production and service. These implications are the result of manufacturing flaws and service impact damage that cannot be ignored.

In the aerospace industry, manufacturers have increasingly moved towards replacing various metallic primary structures with continuous fiber composites. The move to larger and larger composite structures requiring fewer assembly steps has increased the complexity involved in designing, processing and evaluating such structures. The composite elements and larger assemblies in these structures must all account for the effects of defects in the design. This reality is due to the near impossibility of achieving parts with no manufacturing defects, the reality of impact events in service, and the inherent notch sensitive behavior of composites. Defects may or may not be significant depending on the type of defect and the loading environment where these defects occur. Therefore, design values may require built-in conservatism to accommodate the degrading effect of expected defects.

Building on the direct experience of the authors in industry and academia, this book introduces the reader to the latest strategies in the development of a composite structure and the assessment of manufacturing and service defects. The book comprises six major sections.

In Chapters 1 and 2, we discuss introductory material characteristics of composites, such as theoretical constituent behavior, environmental effects and impact sensitivity. These chapters are meant to acquaint the reader with the essentials on composite theory and practice. The latest regulatory requirements are linked to relevant standards associated with the design philosophy and criteria for composite structures. While they may seem more theoretical in nature, these chapters provide the reader with the necessary fundamentals to understand the variability of composite materials and the impact on design.

Chapter 3, the largest chapter of the book, contains the detailed discussion of the various material, manufacturing and service defects that can be encountered in a composite structure. Chapter 4 is a guide to the nondestructive testing techniques that can be used for inspecting and evaluating composite structures with defects. The techniques are evaluated based on their ability to detect a given defect, maturity of approach and whether it is still an emerging technique. Chapter 5 presents guidelines for assessing defects and provides a practical guide for testing strategies and structural analysis of the possible defects. The chapter discusses the properties that can be impacted by these defects and the potential severity. We also present methods for statistical analysis and processing of new experimental effects of defects test data. Finally, the last chapter presents a series of case studies and lessons learned from service experience.

The unconventional nature of this book – by focusing on practical design and defects – will hopefully complement other books on composites by bringing in a healthy dose of reality and skepticism to composite design. We hope it can be a practical resource to students and professionals in the rapidly growing field of composite materials and their structures.

1

Characteristics of Composites

The purpose of this introduction to composites is to review the essential differences of their structural behavior from the more traditional metallic materials. Since the use of composites has become much more commonplace in the past 20 years, and many of the issues are now well known, this review will not be prolonged; instead, it will reference sources in the literature rather than repeat previously documented information.

1.1   Introduction to Behavior

Advanced polymeric composite material systems consist of fibers of high specific strength and stiffness embedded in a resin matrix of significantly lower stiffness and strength. (Note this description is not applicable to ceramic matrix composite materials.) The finished material can be of a laminar or woven structure, with or without core materials. Both the inhomogeneity of the finished structure and its orthotropic structural response are two major characteristics that are not seen in metallic structures to anywhere near the same degree. These structural characteristics lead to the following behavioral differences:

- Composites are notch sensitive, but they shouldn't be considered "brittle." Cracks, holes, and other notch geometries that cause a stress concentration will reduce tension, compression, and shear strength. Matrix-dominated damage types, such as delamination, can grow after development of a "sharp crack," assuming out-of-plane loads can sustain growth.
- Composites are dimensionally stable in-plane, but out-of-plane/through-thickness loads are important because composites are generally anisotropic with the lowest strength coming in out-of-plane tension and shear. The very low thermal expansion can lead to significant residual stresses when composites are combined with metals in hybrid structure. Residual stresses

Composite Structures: Effects of Defects, First Edition.
Rani Elhajjar, Peter Grant and Cindy Ashforth.
© 2019 John Wiley & Sons Ltd. Published 2019 by John Wiley & Sons Ltd.

can also occur in the composite in service or during the cure process due the different coefficients of expansion between the fiber and matrix.

- Composite structures will generally have competing failure modes that will be a function of specific part geometry and may vary with environment.
- Composites are more sensitive to environmental effects and overheating. Most composite matrices absorb moisture in a diffusion process that adds weight. Tension, compression, and shear strength can all be affected by temperature and moisture content.
- Composites typically exhibit flat S-N curves if defects are not present.
- Composite structure must be characterized as a function of laminate layup, materials, thickness, process variables, and individual part geometry.

An unfortunate consequence of these behaviors is that composite characterization is non-standardized and typically proprietary for each composite manufacturer. Composites have high non-recurring costs for product development and certification; therefore, most companies are reluctant to share the composite technology from their investment. In addition, existing composites standards organizations, such as *Composite Materials Handbook-17* (CMH-17) and American Society for Testing and Materials (ASTM) Committee D30, are stymied in their standardization efforts because they are primarily based on volunteer support. In contrast, metal standardization has benefited from governmental actions to accelerate development and ensure a trained workforce during times of need, such as the Second World War. There are no similar driving functions to push composite standardization at this time.

To further describe composite behavior, Table 1.1 provides a general overview of the differences between metallic and composite static strength.

Table 1.1 Differences between metallic and composite structural static strength.

Metal structure	Composite structure
Tensile residual strength is affected by cracks (compression response is typically not).	Tensile and compressive strength are affected by stress concentrations.
Yielding to minimize the effects of small holes and design details that cause stress concentration.	*Localized bearing failure* in reaction to small damage, holes, and design details that cause stress concentrations.
Net section analysis may be used to size static strength for the presence of holes.	Semi-empirical methods or advanced analyses needed to size composite structure with stress concentrations.
Not sized for static strength with cracks or "manufacturing defects."	Static strength sizing with acceptable damage and manufacturing defects.

1.2   Introduction to Composite Analysis

In general, the discrepancy between the fiber and matrix stiffnesses is very significant for most fiber-reinforced composite materials. Typically, the graphite (or carbon fiber) longitudinal Young's modulus is 20 times that of the matrix. For glass fibers, this discrepancy is not as great, and the ratio between the two moduli is typically 3.5. It is well known that these fiber-matrix stiffness discrepancies are difficult to model for an individual ply, and result in increased analytical difficulties of composite structures compared to the analysis of metallic structures. In addition, most practical composite structures consist of multidirectional laminates. The development of laminate structural properties and ply-level stresses further increases the analytical effort compared to that required for homogeneous, isotropic, and metallic structures. Practical (semi-empirical) structural failure prediction methods have been well documented in the literature; examples being presented by Halpin [1], Jones [2], and Jayne and Suddarth [3]. For the purposes of this book, the relationships required to develop multidirectional laminate stiffnesses and lamina (ply)-level stresses and strains for a laminate subjected to in-plane and flexural loading are documented as follows.

The Hooke's law relationships for orthotropic laminae in a plane stress state are:

$$\sigma_1 = Q_{11}\varepsilon_1 + Q_{12}\varepsilon_2$$
$$\sigma_2 = Q_{12}\varepsilon_1 + Q_{22}\varepsilon_2$$
$$\tau_{12} = Q_{66}\gamma_{12}$$

The components of the lamina stiffness matrix (Q) are:

$$Q_{11} = E_{11}/(1 - \nu_{12}\nu_{21})$$
$$Q_{22} = E_{22}/(1 - \nu_{12}\nu_{21})$$
$$Q_{12} = \nu_{21}E_{11}/(1 - \nu_{12}\nu_{21})$$
$$Q_{66} = G_{12}$$

In industrial applications, the load-strain and curvature relationships for the complete multidirectional laminate under applied in-plane loading, are developed using Classical Laminated Plate Theory (CLPT), resulting in the following:

$$\begin{pmatrix} N_x \\ N_y \\ N_{xy} \end{pmatrix} = \begin{pmatrix} A_{11} & A_{12} & A_{16} \\ A_{21} & A_{22} & A_{26} \\ A_{16} & A_{26} & A_{66} \end{pmatrix} \begin{pmatrix} \varepsilon_x^0 \\ \varepsilon_y^0 \\ \gamma_{xy}^0 \end{pmatrix} + \begin{pmatrix} B_{11} & B_{12} & B_{16} \\ B_{21} & B_{22} & B_{26} \\ B_{16} & B_{26} & B_{66} \end{pmatrix} \begin{pmatrix} \kappa_x^0 \\ \kappa_y^0 \\ \kappa_{xy}^0 \end{pmatrix} \tag{1.1}$$

where the superscript "0" denotes mid-plane strains and curvatures.

The matrices A and B are the transformed stiffness matrices (from the lamina axes to the laminate principal axes, "x" and "y") and summated (through the laminate thickness) lamina stiffness matrices (Q). The derivations can be found in Refs. [2, 4]. For the general multidirectional laminate both the A and B matrices are fully populated, which would not be the case for an isotropic material.

In the case of a multidirectional laminate with through the thickness symmetry about the mid-plane (i.e., in thermal equilibrium) the B matrix becomes zero, and an applied system of loads, N, at the mid-plane will not result in any net curvature. In this case, also if the terms A_{16}, and A_{26} in the A matrix are zero, the laminate is said to exhibit zero extensional-shear coupling. This family of laminates are often referred to has being symmetric about their principal (i.e., x and y) axes. Note that this may not be the case for loading directions other than the principal directions (i.e., x and y).

Similarly, the moment-strain relationships can be expressed as follows:

$$
\begin{pmatrix} M_x \\ M_y \\ M_{xy} \end{pmatrix} = \begin{pmatrix} B_{11} & B_{12} & B_{16} \\ B_{21} & B_{22} & B_{26} \\ B_{16} & B_{26} & B_{66} \end{pmatrix} \begin{pmatrix} \varepsilon_x^0 \\ \varepsilon_y^0 \\ \gamma_{xy}^0 \end{pmatrix} + \begin{pmatrix} D_{11} & D_{12} & D_{16} \\ D_{21} & D_{22} & D_{26} \\ D_{16} & D_{26} & D_{66} \end{pmatrix} \begin{pmatrix} \kappa_x \\ \kappa_y \\ \kappa_{xy} \end{pmatrix} \tag{1.2}
$$

Equations (1.1) and (1.2) are commonly combined and expressed as follows:

$$
\begin{bmatrix} N \\ \hline M \end{bmatrix} = \begin{bmatrix} A & B \\ \hline B & D \end{bmatrix} \begin{bmatrix} \varepsilon^0 \\ \hline \kappa \end{bmatrix} \tag{1.3}
$$

In most practical applications, it is normal to assume linear stress-strain behavior at the ply level for the moduli E_{11} and E_{22}. In the case of the latter this assumption is very inaccurate. However, since $E_{11} \gg E_{22}$ the effect of this assumption on the total stiffness of practical fiber-dominated multidirectional laminates is relatively small, and for most practical purposes this assumption is not an issue. Exceptions may occur where it is necessary to determine lamina transverse strains, which are often needed to support the use of lamina-level failure criteria. In those cases, the difficulties in development of valid lamina transverse stress-strain data for both tension and compression loading, can cause insurmountable problems. The use of very low assumed transverse moduli (typically 0.1 × the initial modulus) accounting for matrix damage has been successful [5]. However, the shear modulus, G_{12}, is very nonlinear and can have a significant influence upon the laminate stiffness in some cases, particularly for the (0, ±45) laminate family. For linear analyses, a practical approach is to use a secant modulus in a conservative manner at the appropriate working strain [5].

Equations (1.1)–(1.3) represent the behavior of composite laminates under an applied mechanical loading system. However, in contrast to metals, composites can be extremely sensitive to residual stresses, which can be made more critical by both thermal and absorbed moisture effects. Of the two stresses, thermal stresses are usually the most damaging. Absorbed moisture has a much more detrimental effect upon the matrix stiffness properties (discussed later in Chapter 2).

1.3   Failure and Strength Methodologies

In order to develop the approaches that are used to determine the effects of defects upon the strength of composite structures it is essential to understand the strength methodologies that are used in the initial design phase of the structure in the absence of defects and their relationships with any certification requirements if they exist. In industrial applications, these methodologies tend to vary between industries, particularly in the development of design data (see Chapter 2). In any review of the methods used for the strength estimation of composites structures it is first necessary to outline the state-of-the-art understanding of the separate failure modes that can occur at the lamina level, before proceeding to the prediction of unnotched and notched failure at the multidirectional laminate level. Understanding the behavior at this level will then allow an understanding of sandwich structures and bolted or bonded joints discussed in subsequent chapters.

1.3.1   Lamina Failure Modes and their Influence upon Catastrophic Failure of Multidirectional Laminates

Many attempts have been made to predict lamina failures and their effect upon the catastrophic failure of the complete multidirectional laminate. Use of metals-type yield criteria, such as those by Hill [6] and Hoffman [7] in the late 1960s and 1970s, have always been unsuccessful, mainly due to the severe anisotropic properties of unidirectional fiber-reinforced composites. An attempt to account for this anisotropy at the lamina level was made by Tsai and Wu [8] in a tensor polynomial approach. This approach encompassed all other criteria published up to that date, including maximum stress and maximum strain, and other quadratic interaction criteria. Tsai and Wu made important progress in highlighting the inadequacies of Hill-type interaction criteria. A more up-to-date study utilizing micromechanics methods and progressive damage modeling to account for matrix damage development, and predict failure envelopes, is reported in Ref. [9]. This work also includes a significant historical review. However, evidence of successful application of these methods in design and certification has not been reported to date.

A major program in support of the development of failure criteria and design values for composites military structure was performed at British Aerospace (BAE) in the 1970s [10]. It is believed that this work developed an important understanding of lamina-level strength characteristics and their effect upon catastrophic multidirectional laminate failure. This knowledge is important in both understanding the effects of defects and developing the necessary test methodology for reliable certification. One of the authors of this book (P. Grant) was a lead on this program.

The BAE work resulted in strength design methodology for composite structures and was successfully used on several military programs in the United Kingdom in the 1970s and 1980s. This semi-empirical design methodology was developed from the results of a combined test and analysis program, summarized in Refs. [5, 10]. Interaction criteria, including the Tsai–Wu approach, were found to be inadequate in identifying critical lamina failure modes. The understanding of critical lamina failure modes is important in the development of design values and allowables test programs. The use of large test programs in the absence of a thorough understanding of failure modes can be both expensive and inaccurate. The work at BAE showed that use of pure lamina-level strengths alone, is inadequate in estimating catastrophic failure of the full multidirectional laminate and the influences of adjacent plies and residual stresses must be accounted for. This program identified critical fiber-dominated failure modes in the lamina longitudinal direction. Interaction between adjacent plies was accounted for in the use of in situ lamina strengths developed from cross-ply (0/90) tape laminate data (it is worth noting that the 0/90 configuration is also present in a zero degree fabric ply). In lamina, longitudinal tensile failures, the need to account for matrix stresses was shown to be important, indicating some interaction between matrix stresses and fiber tension strength. This interaction was conservatively accounted for in the cross-ply (i.e., the 0/90 laminate) tension data. The lower bound of longitudinal lamina compression strength was generally found to be an in-plane kink-band mode of instability (see Figure 1.1), which was shown to be strongly dependent upon the matrix shear modulus. The fiber microbuckling failure modes (first studied by Rosen [11]), also shown in Figure 1.1, were not found to be critical – at least in the case of epoxy resin matrices. Whether fiber microbuckling or kink-band is the critical mechanism, it is clear that the resin moduli play an important part in lamina longitudinal compression failure. It is worth noting that the kink-band failure mode may become more critical in structural areas that have fiber wrinkling defects.

Another important result from this work concerned 0 ± 45 laminates loaded in compression in the $0°$ direction, where test data showed that compression strength exceeded that indicated from pure $0°$ strengths. Analysis indicated that the $0°$ laminae exhibited nonlinear stress-strain characteristics after the pure $0°$ strength was exceeded. It was concluded that the $\pm45°$ laminae were providing postbuckling support to the $0°$ laminae. However, the strain levels at failure in these tests were not achieved in laminates that included even small amounts of $90°$ laminae. It was concluded that the use of the pure 0 ± 45 laminate in obtaining strength data should be discouraged, and that at

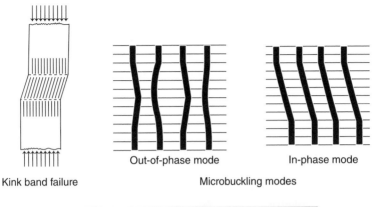

| Kink band failure | Out-of-phase mode | In-phase mode |

Microbuckling modes

Figure 1.1 Fiber instability failure modes [5].

least some 90° plies should always be included to obtain the lower bound strength.

Lamina-level shear strength was developed from tension tests on a ±45° laminate. Interaction criteria between tension and shear, and compression and shear, were developed from multidirectional laminate tests.

It was concluded that strength data from pure lamina tests (from tape material) should not be used in the strength estimation of multidirectional laminates, unless the following effects can be accounted for:

- Inhibition by the fibers of free thermal and moisture matrix deformations.
- Inhibition of lamina Poisson's strains.
- Adjacent laminae provide a crack bridging mechanism for local lamina in the transverse and shear directions.
- Postbuckling support being provided to the 0° laminae from the ±45° plies inhibiting lamina instability, and enabling the compression loaded laminae to continue to carry load beyond initial instability. In this case the effects of any 90° plies must be accounted for.

The current state-of-the art in developing design values tends to vary somewhat from company to company, but in most cases this involves a substantial amount of multidirectional laminate testing. The failure modes are dependent on the directionality of the stresses in each lamina.

In the transverse tension direction, the local stress concentrating effects between fiber and matrix constituents create near-interface failures in a random manner. These localized failures will connect under increased loading and thereby result in separation for a single lamina. In lamina transverse compression failure, matrix shear failure, and/or interfacial separation will develop as the characteristically shear-dominated failure extends across lamina thickness. In the case of in-plane shear of an individual lamina, the highest shear stresses tend to concentrate near the fiber and matrix interfaces and the final shear failure mechanism runs parallel to these interfaces as illustrated.

The significant difference between failure in tension and compression for a specific laminate design compared to a metallic design, is the ability of a ductile metallic alloy (e.g. aluminum 2024 T3) to yield and redistribute the loads. Small damage regions form in highly concentrated stress areas "blunt" the stress concentrations in composites under axial fatigue loading. In contrast many types of metals form a tensile thru crack that can exacerbate fatigue damage effects. Damage zones can actually relieve stress concentration in fiber-dominated laminates.

1.3.2 Design Values and Environmental Sensitivity

Primary structure carbon-epoxy composite materials are typically cured at a temperature of 121° or 177°C (250° or 350°F), whereas the operating temperatures can range from −54° to 82°C (−65 to 180°F). There is a significant thermal mismatch between the fiber and matrix due to unequal coefficients of thermal expansion of the two constituents. Carbon fibers typically exhibit coefficients of effectively zero in their longitudinal direction, whereas the coefficients of matrix materials can be in the order of $36 \times 10^{-6} °C^{-1}$ ($20 \times 10^{-6} °F^{-1}$). As a consequence, and due to the extreme stiffness mismatch between fiber and matrix, the inhibition of matrix contraction by the embedded fibers results in significant matrix thermal tensile stresses at the lower operating temperatures. These stresses have been known to create matrix cracking [5]. Absorbed moisture, caused by the matrix swelling tendency in the presence of moisture, can alleviate the matrix thermal stresses somewhat. Matrix stresses and damage have been known to reduce fiber-direction tensile strength [10].

The most critical effect of absorbed moisture is upon the resin moduli, reduction of which can result in significant degradation of the lamina longitudinal compression strength due to fiber instabilities [5]. Fibers other than carbon can show different types of sensitivities. For example, the aramid (e.g., KEVLAR, DuPont company trademark) fiber will absorb moisture itself, which can create some structural problems. Also, glass fiber composites tend to absorb moisture along the resin-fiber interface, which results in anomalous moisture uptake characteristics.

Evaluation of the effects of moisture can become more critical in the areas of composite structures that have manufacturing or in-service defects. Excessive voids or porosity may absorb moisture to a significantly greater extent than normal. An excellent review of environmental effects is presented in Ref. [4]. Composites also have similar sensitivity to aircraft fluids such as fuels and aviation oil. Moisture uptake values of up to 50% of that of moisture in a typical humid environment are presented in Ref. [4] and are discussed in subsequent chapters. Epoxy based composites have the most sensitivity to moisture and should always be tested to verify the effects. The effects of moisture and other fluids should be evaluated. Section 5.5.4 of the Federal Aviation Administration (FAA) report DOT/FAA/AR-06/10 provides guidance on how this can be accomplished [12].

Due to the multiplicity of failure modes and their different critical environments, it is necessary to develop design values that encompass the complete service environmental envelope. For example, tension failures are normally critical with the material in the low temperature dry condition, whereas compression failures are normally critical with the material having maximum absorbed moisture and at the highest service temperature. Laminate in-plane shear strength may be critical at either extreme, since this can be either compression or tension dominant in the laminae at 45° to the applied shear loading.

Tension failures can be influenced by matrix stresses and matrix damage. As the composite cools down after cure, the matrix contraction is inhibited by the fibers, which in most cases (e.g., as in graphite or carbon fibers) have an almost zero coefficient of expansion in the fiber longitudinal direction. As a result, the matrix experiences tension stresses, which are at their highest at the lowest temperature in the service environment, with the matrix in the "dry" condition. Thermal stresses begin to form at what has been referred to as the fiber-matrix "lock-on" temperature, which is somewhat below the matrix cure temperature. Matrix damage due to thermal stresses alone has been observed at room temperature and below [5]. One effect of absorbed moisture is to create a swelling tendency in the matrix, which can result in some relaxation of matrix tensile stresses. However, moisture can also create matrix damage due to expansion at the freezing point of water. This may be particularly important in laminates with high porosity and/or significant delaminations, and must be addressed in the evaluation of defects that have porosity and higher than normal resin pockets. Repeated freeze-thaw cycles may create environmental fatigue issues in these situations, and this must also be included in any material defect evaluation. Further discussion of these issues and their increased significance in sandwich structures is discussed in Chapter 3.

Compression strength can show significant reduction at higher temperatures. This effect is significantly exacerbated in the presence of absorbed moisture. This is due to the dependency of fiber stability upon matrix moduli. Fiber instability is a well-known contributor to fiber-direction compression strength. The matrix moduli diminish significantly in a nonlinear manner as the temperature approaches the matrix glass transition temperature, *Tg*. Additionally, the presence of absorbed moisture can reduce the *Tg*, resulting in an additional reduction in matrix moduli. A typical curve comparing compression strength for nominally dry ("as received") materials with moisture conditioned materials versus temperature is shown in Figure 1.2 [13]. Metallic materials exhibit significantly less sensitivity to the combined effects of absorbed moisture and temperature, as shown in the typical comparison in Figure 1.3.

In the presence of humid atmosphere or water, carbon-epoxy composites typically absorb moisture and can exhibit weight gains at equilibrium up to 1.5%. This absorption can take place slowly extending over several years, depending upon the laminate thickness. Moisture equilibrium occurs when the moisture content reaches equilibrium with the relative humidity of the surrounding environment and no further change occurs in the material moisture content. This phenomenon is well documented in the literature and a useful review can be found in Ref. [4].

In most practical situations, however, the relative humidity is constantly changing, which can result in a cyclical moisture content in the outer plies accompanied by an approximately constant moisture constant in the inner plies of the laminate. In these cases, a conservative practical approach taken in the development of material properties is to develop design values at the

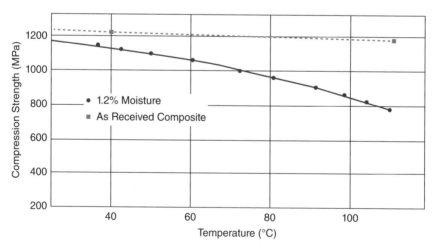

Figure 1.2 Influence of moisture upon the compression strength of unidirectional carbon fiber at varying temperatures.

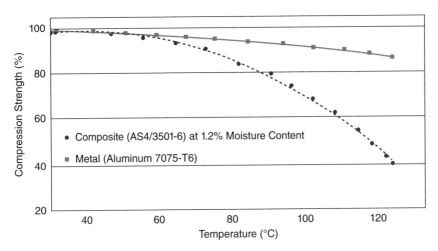

Figure 1.3 Influence of the environment upon composite strength retention as compared to a typical metal.

maximum expected in-service relative humidity. However, this can be overly conservative, in which case materials used in the generation of allowables are conditioned to equilibrium moisture content in an environment representative of a longer term relative humidity condition [14]. The moisture conditioning procedure becomes particularly important in the development of design values for defects, and will be discussed later in Chapter 2.

1.3.3 Design Values for Unnotched Multidirectional Laminates

In general, most structural aerospace applications are designed using notched data. However, there are some cases where unnotched values are used, such as quality control testing and in support of some analytical methods. The approaches vary, and can depend upon the requirements of certifying agencies. The FAA for example, does not specify which allowables to use, but rather lays down the requirements required to derive the statistical design allowables [15]. By contrast, military programs may have specific requirements for generating design values. Outside of aerospace the use of unnotched properties is more common, particularly where safety factors are more than 4.0. Unnotched properties are developed from analytical-experimental based approaches like the BAE approach [5], or from tests on a series of multidirectional laminates, encompassing the full laminate design envelope. In the latter case the number of tests can be prohibitive, and a hybrid approach is often used. The BAE approach used simple cross-ply laminates to develop lamina-level strengths for tape material that accounted for the effects of the multidirectional laminate, along with minimal confirmatory multidirectional laminate data. In considering multidirectional laminates made from fabric material only, the fact that the fabric ply constitutes the actual 0/90 configuration makes the situation much easier. For the hybrid (mixed tape and fabric plies) configuration accurate strength estimation can only be made from tests on the actual multidirectional laminates.

One of the issues with developing unnotched strength data is the use of test configurations that are free from invalid failures. Problems encountered include the notch sensitivity of composites at failure to a degree not encountered in metals, free-edge effects in multidirectional laminates, and the instability issues associated with typical composite compression test specimens. Test specimens with notches in the form of open or filled holes are much easier to test, due to the lack of grip failures and lower failure loads. Generally, "unnotched" failure tends to be initiated by some defect or stress raiser at the macro scale, rather than the micro scale that occurs in metals.

1.3.4 Design Values for Notched Multidirectional Laminates

In advanced fiber-reinforced composites there is no purely analytical method for predicting failure in the region of a severe stress gradient or multiaxial stress condition, such as that which occurs in the presence of holes, free edges, impact damage, and – most importantly for this book – defects. Use of elastic stress concentration factors in the prediction of failures at a filled or open hole in composite laminates can result in an extremely conservative strength. A useful survey of prediction methodology and comparisons with data is presented in Ref. [16]. It is shown that typical notch factors at failure for 6.35 mm (0.25 in) diameter holes in graphite-epoxy composites range from 2.0 to 2.5. Generally, in the case of the lower modulus fibers such as glass and aramid, approximations have in the past proved sufficient mainly due to the relatively nonlinear behavior exhibited in the fiber directions of the laminates composed of these fibers. However, in the case of the advanced intermediate and high modulus fibers used in critical structures and employing lower safety factors, accurate knowledge of the strength of these structures is essential.

In design of composite structures tension and compression notched design values and/or allowables are developed from a database of test coupons having a filled or open hole. (See Table 1.2 for a discussion of design value versus allowable.) Unnotched laminate strengths should not be used for design values because they seldom represent the actual structure. Typical testing methods used are outlined in the ASTM *International* standard practices in Refs. [17–19]. The major part of this database is centered on a hole size commonly found in the structural design. Due to the relatively small coupon widths and lack of consistency (although small) from coupon-to-coupon, correction of the database to infinite-plate conditions is usually found to be necessary. This is normally done using the isotropic width correction factor [20]. This

Table 1.2 Differences between allowables and design values.

Term[a]	Description	Typical use
Allowable	A strength value derived from the statistical reduction of test data from a stable material and laminate fabrication process. The amount of material batches and data required to derive these values is governed by the statistical significance (or basis) needed.	Forms basis of design values. Lower bound strength estimate.
Design value	A strength value used in analysis to compute margins-of-safety. This value most often is based on an allowable value adjusted to account for program design criteria and specific structural conditions.	Depending on allowable basis, used on designs to calculate margin. Can be reduced to account for environment, fasteners, defects, etc.

a) The term "statistical design values" used in 14 CFR 2x.613 and AC25.613-1 is essentially equivalent to the term "allowable."

isotropic relationship is accurate enough for small correction factors which are of the order of 1.03 (e.g., 38.1 mm (1.5 in) wide coupons having a 6.35 mm (0.25 in) diameter hole). Reference [21] discusses the influence of laminate configurations upon this relationship.

Unlike metallic materials, composite stress concentration factors at typical fastener holes at ultimate failure vary with hole size [16], and the design methodology must accommodate this. An analytical approach for hole size correction using a fracture mechanics approach is presented in Ref. [22], and the following equation developed from this reference has been used with some accuracy, at least for initial sizing:

$$\varepsilon_{d1} = \left(\frac{d2}{d1}\right)^{0.3} \times \varepsilon_{d2} \tag{1.4}$$

where ε_{d1} and ε_{d2} are the infinite-plate (i.e., gross) allowable strains for hole diameters, d1 and d2, respectively.

In practice, limited tests on other hole sizes are often performed to support the use, or in lieu of this equation. However, this equation does indicate that the stress concentration at failure increases with hole size until the "large hole situation" is encountered, where in theory failure occurs at the hole perimeter at the unnotched strength, with the strain at that predicted by the elastic stress concentration factor. In aerospace structures a practical example of the "large hole situation" would be a cutout, and in this case, specific design values are normally obtained through testing rather than analysis, often with artificial damage applied near the hole perimeter.

As with bolted joint design, data developed from the notched tests are used as general allowables in the design of structure, with the intent of accounting for unknown damage/stress raisers. However, this approach will not accurately account for many manufacturing and in-service defects.

Experience has shown that the failure mode around a fastener hole can be significantly affected by both the fastener type and installation torque in both tension and compression loading. This is obviously not unexpected in the case of compression loading; however, tension strength can also be significantly influenced by fastener clamp-up by up to 30% (e.g., Ref. [23]).

Initially, in the late 1960s and early 1970s, use of the open hole in design values was believed to account for impact damage. However, as more experience was gained it was found necessary to develop specific data to account for compression strength reductions due to Barely Visible Impact Damage (frequently referred to as BVID), that may occur both in manufacture and in service. Impact damage data developed from coupon tests cannot be extrapolated for accurate use on larger structures; hence point design tests must be performed at the structural element level (see Chapter 5). The types of impact damage that must be accounted for includes both visible and BVID. The latter may occur because of blunt object damage, which can result in significant internal damage with little obvious damage appearing on the surface. The internal damage is often manifested as relatively large delaminations of an area significantly in excess of that appearing on the surface (see Chapter 3) often accompanied by trans-ply resin cracks and minimal fiber breakage. The approach normally taken is to develop design values at the barely visible damage level that can be detected by a visual inspection of the laminate surfaces. In addition, tension strength reductions are also evaluated and compared to the open and/or filled-hole strength.

Additional experience with the types of defects that are now appearing due to the more extensive modern use of composites, and new manufacturing methods, has also shown that the use of the open or filled-hole allowables, will not always accommodate the reduction in both tension and compression strengths that can occur as a result of the defects outlined in Chapter 3.

1.3.5 Material Variability

Fiber dependent strength properties generally exhibit greater variability than metals, where coefficients of variation (CVs) of up to 7% are not uncommon for modulus and can be sometimes higher for strength values in modern carbon fiber/epoxy systems. Highly resin-dependent properties such as out-of-plane shear and interlaminar tension have been known to exhibit CVs of up to 15%. Table 1.3 shows typical comparisons between aluminum and composites obtained from Ref. [24]. The worst case shown is that of the bonded or cocured structure, where the strength is highly resin dependent. Also, the higher variability of composite structure strength is shown to result in lower design

Table 1.3 Comparison of static strength variability of composites and metals.

Material	Static strength variability	Design value/ Mean
	CV, %	
Aluminum	3.5	0.95
Composite – Laminate	6.5	0.89
Composite bonded/cocured	10	0.84

allowables relative to the mean strength, in order to achieve the same A- or B-basis statistical reliability.

1.3.6 Strain-Based Failure Methodology

In contrast to metals, fiber-direction composite failure prediction is in most cases performed using design strength values expressed as strains, both for in-plane and flexural loading. The complete allowables database for fiber-direction strengths is generally based on a failure strain methodology. This methodology is not completely valid and is used more for convenience, since strains do not vary throughout the laminate to the same extent as stresses. The ultimate failure methodology may in the future be based upon ply-level stresses using a validated stress based failure criterion (when one is available). Out-of-plane tension and shear strengths, which are resin-dominated, are predicted using failure stresses.

1.3.7 Composite Fatigue Behavior

Up until fairly recently, there has been a concept that fatigue in composite structures is not a critical issue when developing allowables. In full-scale testing, realistic aircraft (or component) spectra are typically used, but fatigue should also be examined at the coupon level as well. Initially, many of the composite coupon-based fatigue tests were performed under constant amplitude, R = +0.1 loading conditions. This can often be the critical loading condition for metallic structures, but this form of loading is rarely critical for composite structures. Fatigue in composite structures is very dependent on damage growth at both the micro and macro level, and the consequent residual strength reduction at local detail. Consequently, in coupon-based fatigue testing it is important to address the critical loadings for both damage initiation and growth, and residual strength reduction, and these may not be the same conditions. For example, it is now known (e.g., Ref. [25]) that for filled and open hole dominated failures, repeated tension loading will often create extensive damage in and around the hole without reducing – and in some cases even increasing – residual tension strength. However, a significant reduction often occurs in residual compression strength due to the damage created by the tension cycling. In this case it is important to address both tension and compression loading. Hole elongation created by bearing stresses in fatigue loading can occur in composite bolted joints to a degree significantly greater than that in metallic joints (e.g., Ref. [26]). Generally, R = −1.0 cyclic loading will provide a conservative assessment in bolted joints. For impact damage and porosity or waviness type defects, some form of reversed loading condition should always be evaluated.

Composite design values are often based on failure in extreme environmental conditions, such as elevated temperature/wet and low temperature/dry; with the moisture content representative of that nearing the end-of-service lifetime. These conditions may not occur very frequently in the loading spectrum and fatigue testing at these extremes may result in overly conservative data. In aircraft structures a large percentage of the damaging loads occur at temperatures near to room temperature, with the material having moderate moisture content. Fatigue cycling of a full-scale structure is typically performed at nominal environmental conditions, but residual strengths are assessed at environmental extremes. It is important in fatigue cycling that the environmental conditions are representative of typical conditions in the service environmental spectrum and not the extremes. However, residual strength assessments after fatigue cycling should be performed accounting for the environmental extremes.

References

1 Halpin, J.C., Primer on Composite Materials Analysis, (Revised). 1992, CRC Press.
2 Jones, R.M., Mechanics of Composite Materials. 1998, CRC Press.
3 Jayne, B. and S. Suddarth, Matrix-tensor mathematics in orthotropic elasticity, in Orientation Effects in the Mechanical Behavior of Anisotropic Structural Materials. 1966, ASTM International.
4 Springer, G.S., Environmental Effects on Composite Materials. Volume 3. 1988.
5 Sanders, R., E. Edge, and P. Grant, Basic failure mechanisms of laminated composites and related aircraft design implications, in Composite Structures 2. 1983, Springer. pp. 467–485.
6 Hill, R., The Mathematical Theory of Plasticity. Volume 11. 1998, Oxford University Press.
7 Hoffman, O., The brittle strength of orthotropic materials. Journal of Composite Materials, 1967. 1(2): pp. 200–206.

8 Tsai, S.W. and E.M. Wu, A general theory of strength for anisotropic materials. Journal of Composite Materials, 1971. 5(1): pp. 58–80.

9 Levi-Sasson, A., Aboudi, J., Matzenmiller, A., and Haj-Ali, R., Failure envelopes for laminated composites by the parametric HFGMC micromechanical framework. Composite Structures, 2016. 140: pp. 378–389.

10 Sanders, R.C. and I.C. Taig, Final Report on MOD Carbon Fiber Composites, Basic Technology Programme. 1979, British Aerospace, SOR (P) 120 (Warton) September 1979.

11 Fleck, N. and B. Budiansky, Compressive failure of fiber composites due to microbuckling, in Inelastic Deformation of Composite Materials. 1991, Springer. pp. 235–273.

12 Ward, S., W. McCarvill, and J. Tomblin, Guidelines and Recommended Criteria for the Development of a Material Specification for Carbon Fiber/Epoxy Fabric Prepregs. 2007, U.S. Department of Transportation, Federal Aviation Administration: Springfield, VA.

13 Grant, P. and P.L. McConnell. Evaluation and certification of advanced composites for aerospace structures. In Third Canadian Symposium on Aerospace Structures and Materials. 1986, Ontario, Canada.

14 Collings, T., The effect of observed climatic conditions on the moisture equilibrium level of fiber-reinforced plastics. Composites, 1986. 17(1): pp. 33–41.

15 Tomblin, J., Y. Ng, and K.S. Raju, Material Qualification and Equivalency for Polymer Matrix Composite Material Systems: Updated Procedure. 2003, U.S. Department of Transportation, Federal Aviation Administration: Springfield, VA.

16 Curtis, A. and P. Grant, The strength of carbon fiber composite plates with loaded and unloaded holes. Composite Structures, 1984. 2(3): pp. 201–221.

17 International, A., D5766/D5766M-11 Standard Test Method for Open-Hole Tensile Strength of Polymer Matrix Composite Laminates. 2011. ASTM International: West Conshohocken, PA.

18 International, A., D 6484 Standard Test for Open-Hole Compressive Strength of Polymer Matrix Composite Laminates. 2014. ASTM International: West Conshohocken, PA.

19 International, A., D6742/D6742M-17 Standard Practice for Filled Hole Tension and Compression Testing of Polymer Matrix Composite Laminates. 2017. ASTM International: West Conshohocken, PA.

20 Pilkey, W.D. and Pilkey, D.F., Peterson's Stress Concentration Factors. 2008, John Wiley & Sons, Ltd.

21 Gillespie, J.W. and L.A. Carlsson, Influence of finite width on notched laminate strength predictions. Composites Science and Technology, 1988. 32(1): pp. 15–30.

22 Caprino, G., J. Halpin, and L. Nicolais, Fracture mechanics in composite materials. Composites, 1979. 10(4): pp. 223–227.

23 Grant, P. and A. Sawicki. Development of design and analysis methodology for composite bolted joints. in AHS International Specialists' Meeting on Rotorcraft Basic Research. 1991.

24 Whitehead, R. Certification of primary composite aircraft structures. in New Materials and Fatigue Resistant Aircraft Design, Proceedings 14th ICAS Symposium, D.L. Simpson (Ed.), 1987, EMRS Ltd: Warley.

25 Sawicki, A. and P. Minguet. Comparison of fatigue behavior for composite laminates containing open and filled holes. in Sixteenth Technical Conference of the American Society for Composites. 2001.

26 Grant, P., N. Nguyen, and A. Sawicki. Bearing fatigue and hole elongation in composite bolted joints. in Annual Forum, Proceedings of the American Helicopter Society. 1993, American Helicopter Society.

2

Design Methodology and Regulatory Requirements

In many cases, particularly in the aerospace industry, overall design methodology is driven by regulatory requirements. This approach has resulted from the semi-empirical nature of the strength prediction of composite structures, as outlined in Section 1.3. The wind energy industry generally follows the International Electrotechnical Commission (IEC) guidelines with the design basis established independently by companies based on certifiers like DNV-GL [1]. DNV-GL have developed their own design standards/guidelines based on IEC, but including details they themselves judge are relevant. This chapter will generally focus on the aerospace sector, but additional details about the wind sector guidelines are discussed in Ref. [1].

2.1 Regulatory Requirements

In the aerospace industry, regulatory requirements may differ somewhat depending on the certifying agency in question, but are well defined for each program by each of the relevant certifying agencies. The agencies in the USA are separated into military and civil aircraft requirements.

In the US, civil aviation airworthiness requirements are in Title 14, Code of Federal Regulations (14 CFR) parts 21–36. Of the thousands of paragraphs of requirements, there are very few that are unique to composite materials. Most structural requirements are independent of the material system; however, the means of compliance to the regulations may differ based on whether a metallic or composite material is being used. To help satisfy the regulations, the Federal Aviation Administration (FAA) publishes guidance for certification procedures in policy memos and statements and Advisory Circulars (ACs). The top-level guidance for composite structure is AC 20-107B [2]. Many detailed procedures are further described in the Composite Materials Handbook (CMH-17) [3], of which the FAA is the primary funding agent. CMH-17 started as a Military Handbook (Mil-Hdbk-17), which was supported by the military sector – mainly the U.S. Army. Currently, CMH-17 is continually being updated; the

Composite Structures: Effects of Defects, First Edition.
Rani Elhajjar, Peter Grant and Cindy Ashforth.
© 2019 John Wiley & Sons Ltd. Published 2019 by John Wiley & Sons Ltd.

latest version being Revision G [3]. There are excellent reviews on how the FAA interfaces with industry and other national civil aircraft advisory agencies, including the European Aviation Safety Agency (EASA) and Transport Canada Civil Aviation (TCCA), by Ilcewicz et al. [4] and Ashforth et al. [5].

The United States Air Force, Navy (NAVAIR structures), and the Army generally have separate certification requirements for each aircraft and/or rotorcraft program. The airframe requirements are usually controlled separately for each specific program, and the test plans are developed in close collaboration with the certifying agency. Hence the design methodology used also arises from close collaboration with the certifying agency.

In recent years in the aerospace industry the "building block" methodology for structural qualification of composite structures has become the norm. This method is shown schematically in Figure 2.1. In some cases, particularly in the design of space structures, where few replicates of the flight vehicle are

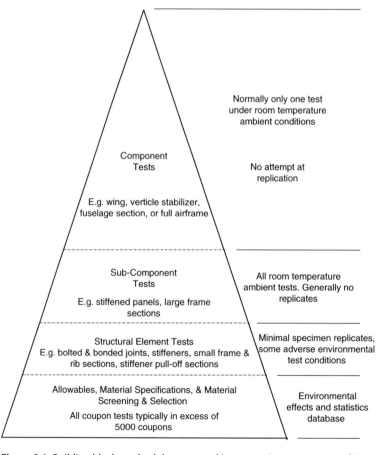

Figure 2.1 Building block methodology as used in composite aerospace applications.

produced, only the top and bottom levels of the building block test plan are performed. The general aviation industry also sometimes utilizes this option, which is considered "certification by test" rather than "certification by test supported by analysis" [2]. In the "certification by test" option, the number of tests in the bottom level is minimized, with often only lamina tests performed, and structural verification is then demonstrated by full-scale tests. This can be a low-cost option when full-scale article can be built quickly and easily, resulting in lower certification cost when compared to developing a full analytical database with validated design allowables. The limitation of this methodology is that all changes require new testing. This includes substantiation of defects. Therefore, this book will focus on the concept of "certification by analysis supported by test" such that a validated analytical model is created and used for the purposes of substantiating defects.

It is worth noting that regulatory requirements in the aerospace industry can be much more rigorous than in most other industries due to both safety requirements and the fact this industry uses a relatively low ultimate factor of safety, typically not in excess of 1.5. It is this, the high cost of full-scale test articles, and the reduction of risk through sub-scale testing, that have all led to the development of the building block method of compliance for composites structures. Other design approaches not using the complete building block, compensate for the lack of such an approach by using factors of safety of more than 4.0 in some cases.

Properties are derived during all levels of the building block. The one-direction refers to the fiber direction, the two-direction is in the lamina plane perpendicular to the one-direction and three-represents the direction normal to the lamina plane. Typically, lamina (ply)-based properties are defined for:

- In-plane
  - Tension, compression – one- and two-directions
  - Shear 12
  - Stress or strain
- Interlaminar
  - Shear stress, 13, 23
  - Short Beam
  - Flatwise tension, compression stress
  - Curved laminate tension stress
  - Fracture toughness
    Mode 1, Mode 2
    Can be a function of the ply orientations adjacent to crack

Laminate strength properties are defined for:

- Multidirectional fiber layups
- Unnotched, open/filled hole, post-impact, large notch

- Tension and compression – stress or strain
- Laminates, sandwich facesheets
- Bending – unnotched, open/filled hole, post-impact
- Bolted joints
  - Bearing stress
  - Bearing/bypass interaction – stress and strain
  - Fastener pull-thru load
- Local buckling, crippling stress
  - Stiffener section elements

Structural Strength properties are:

- Curved laminates (radius details)
- Post-buckled stiffened panel stress
  - With various levels of impacts
- Stiffener pull-off
- Stiffener flange/cap bending, transverse loading
- Beam shear web stress
  - Stable, post-buckled, with/without cutouts
- Sandwich panel edgeband, ramp details
- Bolted joints
- Bonded joints
  - Adhesive, adherend failure modes

2.2 Material and Process Specifications

The development of material specifications is carried out at the lowest level of the building block plan (Figure 2.1). Design values can only be developed for a stable and consistent material. The material specification is established after multiple tests on multiple batches of fiber and resin combinations. Composite properties are also sensitive to manufacturing processes, which must also be controlled by a process specification. The process specification should be validated throughout the building block to ensure applicability and repeatability for lamina coupons through full size structure. In addition, changes in the manufacturing processes must be validated for effects on material properties and structural scaling effects. Generally, resin dominated properties are the most sensitive to changes in the manufacturing processes. However, generalizations must be made with caution. For example, filled-hole tension strength has been known to be extremely sensitive to changes in fiber surface treatment. In addition to the typical design value tests at the coupon level, other tests that are structurally sensitive to the process specification are stiffener pull-off, bonded joints, impacted specimens, modes I and II fracture, short beam shear, sandwich flexure and shear, and so on. Best practices for developing material and process specifications for the aerospace industry, including approval of changes to processes, are described in FAA ACs 23-20 and 20-107B Appendix 3 and report DOT/FAA/AR-03/19. Inadvertent deviations from the process specifications must be addressed as effects-of-defects issues. These are the subject of subsequent chapters in this book.

2.3   Design Methodology

The unique failure modes of composite structures, as discussed in Chapter 1, have a significant influence in the development of the design methodology. In most modern composite structures, the initial design phase begins with the use of in-plane notched allowables. These are often used as a far-field allowable and are normally developed statistically from a large database. In aerospace, unnotched design values are not used for critical (or primary) structures. Even non-critical (or secondary) structures rarely employ use of unnotched design values or allowables. This is currently due to the lack of a test verified analytical method for accurate composite failure analysis in the presence of a steep in-plane stress gradient. However, it should be noted that in non-aerospace structures or where the safety factors are sufficiently high (i.e., typically more than 4.0) unnotched failure criteria may be used for strength prediction. These failure criteria may be like those discussed in Chapter 1. Also, out-of-plane failures cannot be analytically predicted with the accuracy required, particularly if the safety factors are low. This inability to accurately predict composite strength using analytical methods, without a significant test database with specific structural details, has become a critical issue in the strength analysis of the effects-of-defects that can occur in both manufacturing and in service. Going forward, a hybrid analytical/experimental approach may provide the necessary path. An example approach is suggested in the case study shown in Chapter 6 using a coupled experimental/analytical approach for predicting the failure of a laminate with fiber waviness. Such an approach may help fill reduce the amount of testing required.

As the design process progresses and stress analysis of structural details is performed, more specific design values are required. Typical cases are bolted and bonded joints, structural details that occur in stiffened panels, frames, ribs, sandwich panels, and so on. Design values may also be generated to account for manufacturing defects and impact damages by performing tests on typical structural details that have barely visible impact damage (BVID). Often it is necessary to test larger structures such as multi-stiffened panels, particularly to substantiate larger categories of damage as described by the FAA in AC 20-107B [2]. Test data from damaged specimens at the coupon level cannot be accurately scaled up to the structural detail level. In some defect cases, the coupon can provide a conservative result. However, the design of structures with large damage can only be substantiated by tests on the actual structural parts.

Structural design details that cause stress concentrations will lower the in-plane and out-of-plane strength of composite structures. Although composites do not have a "yielding behavior" they do have "strain softening," which is highly dependent on the specific materials, layups and ply stacking sequences used for a given design. Often, semi-empirical failure criteria are applied to predict the effects of stress concentrations.

Damage, defects, and other manufacturing discrepancies also lead to in-plane and out-of-plane stress concentrations that must be considered for static strength. One additional problem with such defects, is defining accurate damage metrics to represent the damage or defect in a structural analysis. Often, conservative assumptions are used, such as assuming the full impact damage area is an open hole or wrinkled plies have greatly reduced local stiffness, in order to come up with some estimates of the strength. Significant amounts of test data are usually still needed to prove the assumptions are conservative for each damage or defect type.

- Base material properties are important to quantifying variability, environmental effects, and moduli, but have limited use in predicting static strength
- Composite structural failure usually initiates at local stress concentrations (in-plane or out-of-plane) caused by design detail, damage, or manufacturing flaws
- Semi-empirical engineering approaches are typically used to address the many factors that localize damage and affect static strength
- Analysis and test iterations between the various levels of study should be anticipated in developing a complete substantiation of static strength
- All details, which cause local stress concentration, should be understood to avoid premature failures in component tests

It is important to realize that analysis assumptions are often generated to ensure the failure prediction will always be well below the test result, in order to conservatively address the regulations associated with material and manufacturing fabrication and assembly variations. One example of such an analysis assumption is a spreadsheet of the loads distributed to individual wing stringer stations that add up to loads higher than the total in order to ensure a conservative analysis, independent of variations possible in structural assembly.

2.4 Design Values for Notched Multidirectional Laminates

These are developed based on filled and/or open-hole coupon strengths that are often considered as allowables, and are typically required to be developed from a statistically significant database. The hole size is normally taken as that used in typical bolted joints that commonly occur in the structure. Hole-size correction factors to account other hole sizes outside the database are often required and can be developed from limited tests or analysis (see Chapter 1). However analytical methods for hole-size correction are also available in the literature (e.g., Ref. [6]). Semi-Empirical approaches have been used for developing notched strengths from unnotched strengths. A summary of these is presented by Curtis and Grant [7]. However, the notched design value database in airframe structures is more commonly developed from a large database of filled and open-hole strengths. Uses of the semi-empirical approaches are often much more economical, but may have to be substantiated by more complex tests higher in the building block test plan (Figure 2.1). This semi-empirical approach may be adequate for non-aerospace structures where the safety factors employed are much higher.

In structures with a wide range of laminate configurations, the traditional carpet plot method of representing allowables encompassing the complete range of laminates in the structure is frequently used. However, this approach results in many tests needed to develop statistical allowables at either the A- or B-basis level. Also, the carpet plot method assumes that failure strength is a continuous function between laminate configurations, and does not recognize the discontinuous nature of composite failure modes. Due to the large number of laminate configurations tested this may not be a critical issue, but a rigorous quantification of the effects of the various environmental conditions may be missing. An example of the carpet plot method is shown in Figure 2.2.

As compared with the carpet plot method the number of tests was significantly reduced by using limited unnotched data in conjunction with failure criteria [8] and test validated stress concentration factors in work performed at British Aerospace (BAE), in the UK (United Kingdom) in the late 1970s and early 1980s. In most practical laminates the test concentration factors were significantly less than those predicted by the orthotropic elastic stress concentration factors (i.e., Eq. 2.1). Use of the elastic factor in conjunction with unnotched strength data would have in fact have resulted in undue conservatism. The BAE method was accepted by the UK and European military aircraft certification authorities, but there is no evidence of this method being used in the USA.

In lieu of the traditional carpet plot, method, data, and allowables have also been represented as an approximate continuous function developed from the orthotropic elastic stress concentration factor, "K_t," which is expressed in Eq. (5.1) as follows:

$$K_t = 1 + \sqrt{2\left(\sqrt{\frac{E_x}{E_y}} - v_{xy}\right) + \frac{E_x}{G_{xy}}} \qquad (2.1)$$

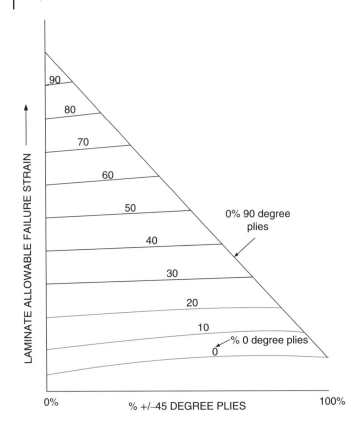

Figure 2.2 Typical carpet plot for allowable failure strains.

This relationship was developed by Lekhnitski [9]. This approach was used on the V-22 Osprey Tilt Rotor program and significantly reduced the number of tests as compared with the carpet plot method. This approach was certified under the NAVAIR structures department, and is summarized in Refs. [10, 11]. On the V-22 program "K_t" was expressed in terms of a laminate configuration parameter known as AML (angle minus loaded: angle plies minus 0° or loaded plies) as follows:

$AML = \%45°$ plies $- \%0°$ plies [10].

"K_t" is shown represented in both the carpet plot format in Figure 2.3 and in terms of the AML parameter in Figure 2.4.

As shown in Figure 2.4 the AML representation for "K_t" almost reduces to a continuous function. The extreme parts of the curves lying off the main trends are in fact representative of laminates that are not typically used in structural applications. The laminate configurations used on the V-22 airframe were from the (0°/±45°/90°) family [10]. It is worth noting that for an isotropic

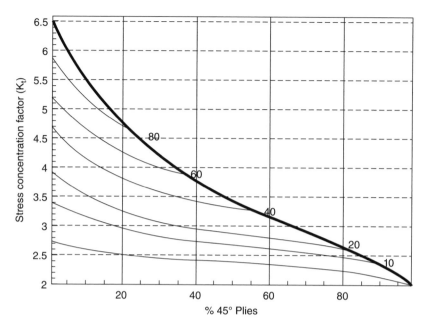

Figure 2.3 Typical orthotropic elastic stress concentration factors for multidirectional laminates.

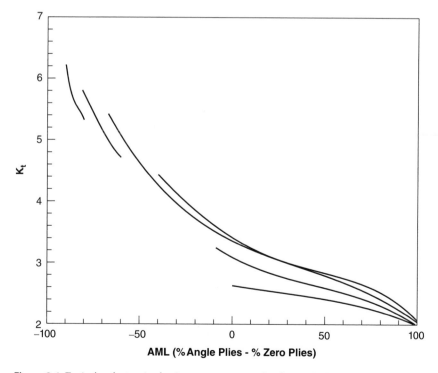

Figure 2.4 Typical orthotropic elastic stress concentration factors in Figure 2.3 expressed in terms of AML.

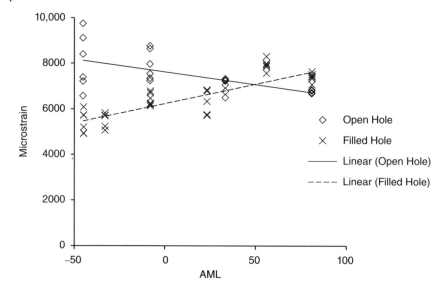

Figure 2.5 Use of the AML parameter (tension properties).

material and for the quasi-isotropic laminate configuration, Eq. (2.1) reduces to a "K_t" of 3.0. The "K_ts" for the orthotropic structural laminates represented in Figure 2.3 range from 2.0 to 7.0. Again, it should be noted as stated in Chapter 1 (Section 1.2) and here that these high values are not normally realized at failure, where typical failure notch factors for graphite/epoxy composites range from 2.0 to 2.5.

Application of the use of the AML parameter is shown in the example in Figure 2.5, which is reproduced from Ref. [10]. This figure shows open and filled-hole failure strain data plotted against the AML parameter. Regression lines are shown through the filled and open-hole data. In most cases in this plot for design allowable purposes the filled-hole line is critical. However, both the filled and open-hole situation should be evaluated. The filled-hole data will also vary with different degrees of fastener clampup. Generally, the design allowable envelope is the minimum of the two regression lines shown in Figure 2.5. As explained in Ref. [11] this approach was used to represent and evaluate effect-of-defects data from tests on tow-placed laminates. The effects-of-defects data were represented as a function of the AML parameter.

2.5 Design Values for Bolted Joints

The design of bolted joints in composite structures is a somewhat controversial topic, and methodologies can vary throughout the industry. Use of traditional metals methods generally prove inadequate when applied to composite joints. Reference [10] presents a methodology developed for use on rotorcraft structure, and is based on methods used on certification of the V-22 Osprey program. Although the authors of this book do not claim that this represents any form of industry standard, this is a work that presents the major issues involved. The following is an overview of methodologies used.

The major issues center on the fastener load distributions, identification of failure modes, and the interaction between these modes. These are important issues in the evaluation of the effect-of-defects upon the strength of composite bolted joints. The strength estimation of a composite joint is performed in two stages as follows:

1) Determination of fastener loading, and
2) Failure analysis of a loaded hole in a general stress field.

2.5.1 Determination of Fastener Loading

The assumption of even load sharing between fasteners in a complex joint is inadequate for accurate strength analysis of composite joints, and some estimation of load peaking is necessary. As stated in [10], use of an elastic methodology in estimating individual fastener loading may lead to conservatism, since local yielding due to bearing and fastener bending will alleviate severe fastener load concentrations, particularly at joint loading near to and at failure. At the time of publication of Ref. [10], methods using fastener and laminate flexibilities to alleviate load peaking have been reported in practical use. However, a comprehensive analytical account of such issues as friction between mating surfaces local to the fasteners, fastener clampup, and extreme bearing deformation may not yet be practicable. Friction effects may be reduced due to fatigue loading. The effects of the combination of moisture absorption and fatigue upon fastener clampup (and hence friction) is a complex topic. It is certain that a purely elastic approach in the estimation of fastener loading would be too conservative, and an approach somewhere between this and an even load distribution is necessary. For these reasons, design methodology of composite bolted joints often needs substantiation by tests.

2.5.2 Failure Analysis at a Loaded and Unloaded Hole

The basic failure modes at unloaded and loaded holes can be separated as follows:
1) Pure bypass failure in tension or compression. This occurs when there is no load reacted at the hole, that is an unloaded hole.
2) Pure bearing. This occurs due to bearing stresses alone.
3) Bearing–bypass interaction; that is, between (1) and (2).

In a multi-fastener bolted joint, the fasteners holes are subjected to bearing loads and loads that bypass the hole. Bypass failure normally occurs in the tangential fibers at or near to the hole perimeter. This is more obvious in tension failures than in compression, since compression failures are dominated by fiber/ply instability and are also complicated by load transfer through the actual fastener [7]. Pure bypass failure at zero bearing stress corresponds with the basic design values for filled and open holes. In compression loading this is somewhat more complicated, due to the fastener effect. Conservatism in the design criteria may lead to the use of open-hole values in compression loading.

Pure bearing data is normally developed from a tension test on a single fastener joint. However, this is not a true pure bearing test, since some bypass stresses exist on the hole perimeter. The only true test is one loaded in compression. However, the tension test is much easier to perform, will normally yield a conservative estimate of bearing strength, and is the test most used. The catastrophic failure modes in the tension bearing test are extremely sensitive to hole edge and end distance and do not represent true bearing strength. These modes normally occur after hole elongation due to bearing damage. This sensitivity to hole location make the tension test very useful for interrogating hole and hole-location defects; in particular, short end and edge distances. Large bearing deformations often occur prior to catastrophic failure and for design purposes a deformation based failure criterion is often used.

An excellent review of bearing test methods is presented in the American Society for Testing and Materials (ASTM) *International* standard D5961/D5961M [12], which is also consistent with the recommendations of CMH-17. Bearing damage is extremely sensitive to environmental effects and friction between the mating surfaces. The critical environment is at high temperature with the adherends in the wet condition. Typical data from Ref. [10], showing the effects of fastener clampup upon bearing design values is presented in Figure 2.6.

This figure compares the effects of low and high fastener clampup (i.e., friction). The data is represented in terms of the percentage of ±45° plies in the multidirectional laminate. All laminates in this database also included 0° and 90° plies. With very low percentages of ±45° plies, typically below 25%, the failure mode tended to be that of shearout, and these cases the data exhibited high variability. This is normally exacerbated by high stratification of the 0° plies. For design purposes, limitations on ply stratification are normally necessary. Damage or defects in highly stratified plies orientated at zero degrees to the principal loading direction can lead to significant strength degradation in both bearing and other loading modes. A similar method of representing bearing failure as a function of laminate configurations is suggested by Collings in Ref. [13]. Interestingly, Collings used compression bearing data in support of his analysis. Collings also generated bearing strength data as a function "lateral constraint" (i.e., fastener clampup). The data is reproduced in Figure 2.7.

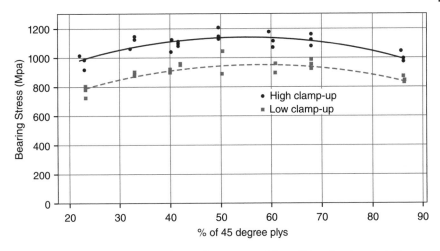

Figure 2.6 Bearing data for the 82°C (180°F)/wet material condition showing the influence of fastener clampup.

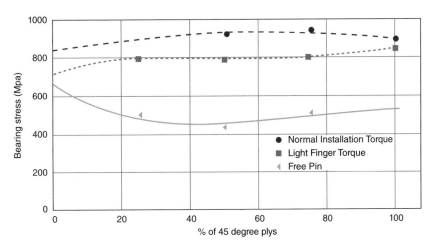

Figure 2.7 The effects of lateral constraint on the bearing strength of HTS/914 laminates. Source: From Ref. [13].

Three degrees of fastener clampup are shown ranging from "free pin" to normal installation torque. Fastener clampup is shown to be an important issue for the topic of this book, since incorrect installation torque and/or its un-accounted for degradation in service should be considered to be defects. In reviewing failure stresses in Figure 2.7, it should be remembered that compression bearing data is normally significantly higher than data obtained from a tension test. Bearing design values are normally obtained from the conservative tension test.

Use of tension bearing design values often under predict structural failures in compression dominated bolted joints, where lateral constraint tends to inhibit fiber and lamina brooming that commonly occurs in compression failures of composite laminates.

The interaction between bearing and bypass stresses is an important issue in the design of composite joints. This occurs to an extent in static failure not seen in metallic joints. The bearing–bypass interaction diagram is shown typically in Figure 2.8. At high bearing stresses extreme scatter has been observed in failure stresses [10]. On the V-22 program this led to the use of a bearing cut-off design value (Figure 2.8), essentially removing the variable data, and possibly also accommodating bearing fatigue effects. A similar effect may be obtained by using a deformation criterion for bearing strength. Various semi-empirical methods for predicting failures in fibers tangential to the hole circumference are summarized in Ref. [7]. Reference [10] reports on use of the Bolted Joint

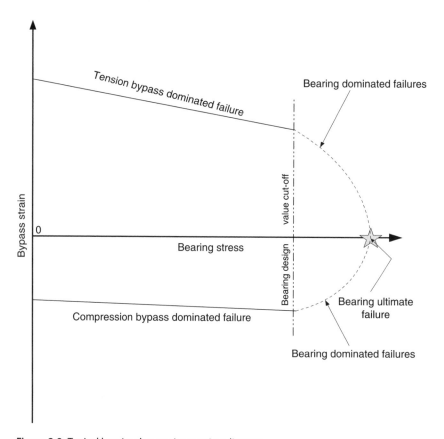

Figure 2.8 Typical bearing-bypass interaction diagram.

Stress Field Model (BJSFM, [14]) combining stresses due to bypass loading and applied bearing to predict failures in the bypass dominated regime (Figure 2.8) using the Whitney–Nuismer Point Stress failure criterion [15]. The interaction diagram then becomes two straight lines for bypass loading and one straight line for the bearing cut-off value. Design criteria may dictate the choice of fastener and its clampup, or open-hole conservatism in the compression bypass dominated regime.

2.6 Design Values for Bonded Joints and Bondlines

The development in the use of adhesives as a means of joining composite structures has been much slower than was anticipated in the 1970s. Originally (in the early 1970s), bonding was considered to be preferable to the use of fasteners due to its lack of stress concentrations at the fastener, and the relatively poor composite bearing performance at the fastener load introduction. However, bonded joints tend to suffer due to poor peel strength at the bondline extremities and a significant sensitivity to bondline quality. These factors along with inadequate bondline inspection capabilities have so far inhibited the use of bonds as a primary means of load transfer in commercial transport category aircraft, which experience high out-of-plane loading. Also the strength of bonded joints can be extremely sensitive to defects and damage in the bondline. The use of small fasteners ("chicken fasteners") through the bondline will alleviate the peel problems, but will not always prevent non-catastrophic peeling at the joint extremities before load transfer reaches these fasteners. It is worth noting that bondlines are much stiffer (due to the small adhesive thickness) in load transfer than the fasteners, and, as a result, the adhesive layer tends to pick up load prior to the fasteners. The ultimate strength of a bolted-bonded joint may not be much greater than that of the fasteners alone. Bonding is still used, however, in the cobonding of stiffeners and other parts in commercial transport category aircraft structures, and applications may exist in non-aerospace structures where safety factors are much higher. Bonded repairs are also common in both aerospace and non-aerospace applications.

The development of design values in bonded joints has mainly concentrated on shear strength. In joints that are sensitive to out-of-plane peel failures, certification must be achieved through testing of the actual structural details. Past design of bonded-metal structures has been achieved by neglecting stress concentrations and employing relatively low adhesive stress allowables, often developed from tests on short (typically 1 to 2 cm long) metallic joints. For the design and assessment of defects in highly loaded joints in a primary structure, accurate knowledge of the adhesive stress distribution and a better understanding of failure is essential.

Modern structural adhesives exhibit significant nonlinearity and high failure strain, particularly in shear loading. The shear strain at failure can be in

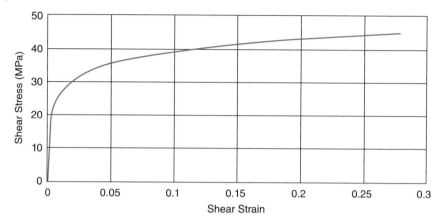

Figure 2.9 Typical structural adhesive shear-strain response, FM300K.

excess of 30%. A typical shear stress–strain curve is shown in Figure 2.9 for the FM300K adhesive (Cytec Industries Inc., Tempe, Arizona USA). This is extremely beneficial in alleviating stress concentrations. Hence any design methodology must account for the true adhesive stress–strain characteristics. In Ref. [16] a closed-form method of analysis for predicting both shear and peel stresses is reported based on the work of P. Grant and I.C. Taig of BAe [17, 18]. In order to develop interaction between peel and shear stresses the analysis allows the joint adherends to deform out-of-plane in a nonlinear manner. However, in the combined shear and peel analysis the adhesive stress–strain characteristics can only be modeled as a linear material. This analysis was an extension of original work by Goland and Reissner [19] to make it applicable to multi-step lapped joints. This analysis also showed the existence of peel stresses in symmetric joints with no applied bending moment. Advanced methods for adhesive shear stresses can account for the full nonlinear behavior of the adhesive in shear loading only. An elastic–plastic model of the adhesive shear stress–strain characteristics was developed by Hart-Smith [20] and used on the PABST (Primary Adhesively Bonded Structure Technology) program.

However, the work reported in Ref. [17] utilizes a model of the adhesive shear stress–strain curve, which closer represents the actual shear curve. Here the nonlinear shear lag analysis utilized a finite difference method of modeling the shear stress distribution along the length of the joint. The importance of joint length in developing joint strength data for a nonlinear adhesive of high failure strain in shear is shown in Figures 2.10 and 2.11. Figure 2.10 compares the shear stress distribution in the 1 cm/0.5 in. joint lap length with that in the 5 cm/2.0 in. lap length. Figure 2.11 shows the effect of adhesive shear failure strain and lap length for a single lap joint. It is important to note the lack of

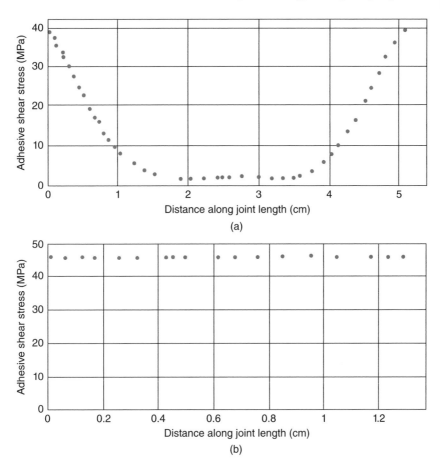

Figure 2.10 Comparison of shear stress distributions of long (a) and short (b) bonded joints with 0.3 mm (0.012 in.) thick adhesive and 2 mm (0.079 in.) thick aluminum adherends. Source: From Ref. [16], fig. 11. (a) Adhesive shear stress distribution in a long joint. (b) Shear stress distribution in a short joint .

sensitivity of load capability at short joint lengths to adhesive failure strain. At the very short lap lengths (e.g., less than 1 cm/0.5 in.) the load can show very little sensitivity to adhesive failure strain, that is, for failure strains of 25 and 7.5% the predicted failure loads are almost the same for the short joints. This indicates that lap lengths in this case should be in excess of 3 cm (1.5 in.) to establish strength data. In any investigations of the effects of adhesive quality and defects upon adhesive shear strength it is important to test joints of sufficient length. Short joints of insufficient lengths may give inaccurate assessments of joint shear strength in structural applications, particularly in the presence of bond-line defects.

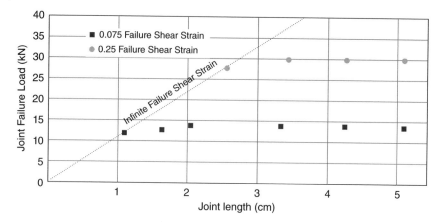

Figure 2.11 Typical variation of bonded joint strength with overlap length and adhesive failure strain in shear for a single lap joint. Source: From Ref. [16], fig. 10. (Strengths shown are predicted based on shear failure strain) .

2.7 Design Values for Sandwich Structure

The approach used for sandwich structures with composite facesheets and a metallic or non-metallic core normally involves developing individual facesheet allowables, and using supplier developed core properties. The core materials may have continuous (e.g., balsa wood and foams) or discontinuous (i.e., honeycomb) facesheet bonding surfaces. Sandwich structures may be designed to react both in-plane loading and flexural loading. For the pure flexure case design values are often developed from a pure flexure test. The ASTM subcommittee D30.09 is a useful resource in the development of standard test methods, practices, guides, and terminology in the area of sandwich construction.

2.7.1 Facesheet Tension Design Values

Use of the normal filled or open-hole values is in most cases adequate. Although in non-primary structure and/or where factors of safety are high (e.g., in excess of 3.0), design values approaching the unnotched values may be used.

2.7.2 Facesheet Compression Design Values

The failure modes of compression loaded sandwich structure make the use of normal solid-laminate design values in most cases inadequate. As a result, compression design values are normally developed or substantiated with pure compression tests on the actual sandwich structure. The integrity of facesheet compression strength can be seriously reduced by an inadequate core-facesheet bondline. Local damage or defects in this can only be addressed by an edge-wise compression test of the sandwich structure. A defective core splice should also be addressed by this test.

2.7.3 Sandwich Flexural Design Values

A sandwich structure under flexural loading can in many cases be designed conservatively using separate facesheet and core properties, without resorting to a flexure test. Supplier-used out-of-plane core shear strengths should be substantiated with design specific shear tests. However, the pure flexure tests will remove any conservatism, but general use of the test results must be performed with some caution, since they may only apply only to the specific design and loading condition tested. Defective core splices must also be addressed using a 3 or 4-point flexure test by ensuring that the spliced section is reacting the design shear loads.

2.7.4 Out-of-Plane Loading

Critical loading conditions are flatwise tension and compression. The former should not be a primary loading condition but the property may be needed in some cases. Compression strength is often needed for load bearing applications. Core compression and/or crushing properties are normally supplied in vendor data. The ASTM tests C297 and C365 should be accessed for accurate core flatwise tension and compression. Core out-of-plane shear strength is often developed from a conservative estimate of vendor data; however the core shear strength is better developed from a flexure or plate shear test.

2.8 Statistical Allowables

In aerospace composite structures B-basis allowables are normally considered acceptable by all the certifying agencies and are used in preference to A-basis allowables, which have traditionally been used in the past for metallic structures. In the USA, these allowables were originally defined in Mil-Hdbk-5 as follows:

- The A-basis value is that value above which 99% of the population of the values is expected to fall with a confidence of 95%.
- The B-basis value is that value above which 90% of the population of the values is expected to fall with a confidence of 95%.

Due to the variability of composite strength data, the use of A-basis values tends to lead to significantly lower allowables than those resulting from B-basis data analysis. However, B-basis values should only be applied to redundant load-path structures. A-basis values should be used for single load-path structures. Industries that can accommodate the more conservative approach may use A-basis allowables as the general design values.

In the aerospace industry, a deterministic value known as the "S" basis value is sometimes used as a conservative allowable. This is a value that can be obtained from material acceptance tests. "Any material when test sampled is rejected if any of its properties when tested fall below the established S-value" (from CMH-17, Vol. 1, ch. 8, [3]).

Due to the fact that composites often exhibit significant batch-to-batch variability it is normally necessary to develop statistical design values from several batches of material. Data sets from this type of material are defined in CMH-17 [3] as "structured datasets." Generally, the use of at least five batches of material is required. For instance, the use of the ANOVA (Analysis Of Variance) method is not recommended in CMH-17 for use on datasets where less than five batches are included.

As an example for a dataset having a normal distribution the A- and B-basis design values are given by:

$$\text{A-basis design value} = M_p - k_a \times s \qquad (2.2)$$

$$\text{B-basis design value} = M_p - k_b \times s \qquad (2.3)$$

where "M_p" is the population mean, "s" is the population standard deviation, and k_a and k_b are the one sided tolerance limit factors. The tolerance limit factors can be obtained from tables 8.5.10 and 11 in Volume 1 of CMH-17. These decrease with an increase in the number of data points samples (test samples). For an example, see Table 2.1 that shows the "B" and "A"-basis allowables for a varying number of data points in a population with a mean of 344.9 MPa (50 ksi) and coefficients of variation of 7 and 15%. A coefficient of variation of 7% is typical for a fiber dominated design value, whereas 15% is typical for a resin dominated design value (see Section 1.2.5), which may also be typical for an effects-of-defects statistically developed design value. It can be seen that there is a significant reduction in allowables with a reduction in the number of data points and an increase in the population coefficient of variation.

The assumption of normality in the database, which is often made, may not be true. CMH-17 has developed an involved process that tests for normality, and other models such as lognormal or two-parameter Weibull may be used. Statistical analysis codes are available from CMH-17. The reader is advised to consult Volume 1, ch. 8 of Ref. [3] for up-to-date information on guidelines for commercial aircraft; in particular, on the ANOVA method. In the past, military programs have assumed normality, with consent of their regulatory agencies.

Table 2.1 Typical allowables for a normal distribution, with a mean of 344.9 MPa (50 ksi).

No of data points	Coefficient of variation, CV %	Standard deviation MPa	k_b	B allowable MPa	k_a	A allowable MPa
Infinite	7	24.14	1.282	313.9	2.326	288.7
30	7	24.14	1.778	301.9	3.064	270.9
10	7	24.14	2.355	288.0	3.981	248.7
Infinite	15	51.72	1.282	278.5	2.326	224.5
30	15	51.72	1.778	252.9	3.064	186.3
10	15	51.72	2.355	223.0	3.981	138.9

2.9 Simulation of Temperature and Moisture Content

Realistic and accurate simulation of environmental conditions for the evaluation of the effects-of-defects upon composite structure strength and stiffness is extremely important, mainly due to the fact that most defects degrade resin dominated properties.

For design values, the extremes of the service temperatures are normally used. Design value data is also obtained at room temperature principally for the evaluation of the more complex structural tests that are performed at room temperature. However, moisture simulation for structural analysis and in test specimens is a more difficult proposition.

The mechanics of moisture absorption and its relation to design values was discussed in Section 1.3.2. In most practical situations structures are exposed to a cyclic environment and merely assuming the worst-case service relative humidity (RH) can be too conservative. Collings [21] has suggested for aircraft that the use of 84% RH will represent worst-case world-wide environmental conditions. Interestingly, CMH-17 have independently arrived at 85% RH has being conservatively suitable for commercial aircraft (vol. 1, ch. 2 of Ref. [3]). However, in the case of a known specific environment with RH in excess of 85% or lower, this known RH may be used. This may tend to occur in non-aerospace applications. Test specimens are generally conditioned until moisture equilibrium is achieved. This can be somewhat of an approximation of real world situations, due to the fact that in-service conditions can be cyclic, which affects the outer plies in the laminate resulting in fluctuations in the moisture content in these plies (Ref. [22] ch. 6). The ASTM *International* test method, D5229 [23] is an excellent resource for moisture conditioning and testing guidelines for polymer matrix composite materials.

The effects of low temperature with the material in both the wet and dry condition can be structurally degrading in the presence of resin pockets and high porosity. In the dry condition at low temperature resin thermal stresses can be significant (Section 1.1). The effect of absorbed moisture may relieve these stresses, but moisture expansion when freezing in resin pockets and high porosity can create damage; hence, both the dry and wet conditions must be addressed in test programs. The possible fatigue damage due to the freeze-thaw effect in environmental cycling must also be considered.

References

1 GL, D., Rotor blades for wind turbines, in DNVGL-ST-0376. 2015, DNV-GL: Bærum, Akershus, Norway.

2 FAA, Composite aircraft structure, in FAA Advisory Circular AC 20-107B. 2009, Federal Aviation Administration.

3 SAE, Composite Materials Handbook, CMH-17-1G. Published by SAE International, ISBN of 978-0-7680-7823-7.

4 Ilcewicz, L. and B. Murphy. Safety & certification initiatives for composite airframe structure. in 46th AIAA/ASME/ASCE/AHS/ASC Structures, Structural Dynamics and Materials Conference. 2005.

5 Ashforth, C., L. Ilcewicz, and R. Jones. Industry and regulatory interface in developing composite airframe certification guidance. In 29th Annual Technical Conference of the American Society for Composites 2014. 2014, La Jolla, CA, USA: DEStech Publications, Inc.

6 Grant, P. and A. Sawicki. Development of design and analysis methodology for composite bolted joints. In AHS International Specialists' Meeting on Rotorcraft Basic Research. 1991.

7 Curtis, A. and P. Grant, The strength of carbon fibre composite plates with loaded and unloaded holes. Composite Structures, 1984. 2(3): pp. 201–221.

8 Sanders, R., E. Edge, and P. Grant, Basic failure mechanisms of laminated composites and related aircraft design implications, In Composite Structures, 2. 1983, Springer. pp. 467–485.

9 Lekhnitskii, S.G., Anisotropic Plates. 1968, DTIC Document.

10 Sawicki, A. and P. Minguet. Comparison of fatigue behavior for composite laminates containing open and filled holes. In Sixteenth Technical Conference of the American Society for Composites, 2001.

11 Sawicki, A., E. Schulze, L. Fitzwater, and Harris, K. Structural qualification of V-22 EMD tow-placed Aft Fuselage. in Annual Forum Proceedings-American Helicopter Society. 1995, American Helicopter Society.

12 ASTM, D5691/D5961M-13 Standard Test Method for Bearing Response of Polymer Matrix Composite Laminates. 2013. ASTM International: West Conshohocken, PA.

13 Collings, T., On the bearing strengths of CFRP laminates. Composites, 1982. 13(3): pp. 241–252.

14 Garbo, S.P. and J. Ogonowski, Effect of Variances and Manufacturing Tolerances on the Design Strength and Life of Mechanically Fastened Composite Joints. Volume 1. Methodology Development and Data Evaluation. 1981, DTIC Document.

15 Whitney, J.M. and R. Nuismer, Stress fracture criteria for laminated composites containing stress concentrations. Journal of Composite Materials, 1974. 8(3): pp. 253–265.

16 Grant, P. and J.A. Lewington, Some important considerations in the design of composite bonded joints in aerospace structures. Canadian Aeronautics and Space Journal, 1987. 33(2): pp. 91–98.

17 Grant, P. Analysis of adhesive stresses in bonded joints. In Symposium: Jointing in Fibre Reinforced Plastics. Imperial College, London, IPC Science and Technology Press Ltd. 1978.

18 Adams, R.D., J. Comyn, and W.C. Wake, Structural Adhesive Joints in Engineering. 1997: Springer Science & Business Media.

19 Goland, M. and E. Reissner, The stresses in cemented joints. J. Applied Mechanics, 1944. 11.

20 Hart-Smith, L.J., Adhesive-Bonded Single-Lap Joints. 1973, Citeseer.

21 Collings, T., The effect of observed climatic conditions on the moisture equilibrium level of fibre-reinforced plastics. Composites, 1986. 17(1): pp. 33–41.

22 Springer, G.S., Environmental Effects on Composite Materials. Volume 3. 1988: Elsevier.

23 ASTM, D5229/D5229M-14 Standard Test Method for Moisture Absorption Properties and Equilibrium Conditioning of Polymer Matrix Composite Materials. 2014. ASTM International: West Conshohocken, PA.

3

Material, Manufacturing, and Service Defects

3.1 Introduction

The terms flaws, defects, and damage can be used synonymously. Sometimes the term "flaw" is meant to refer to items that occur *during* the manufacture of the part or assembly (e.g., wrinkles, porosity, and delamination), and "damage" is introduced *after* the part is cured or assembled – either while still in the factory or after delivery (such as by an impact).

All composite parts have some level of flaws inherent in their manufacture. If nothing else, it is essentially impossible to manufacture a part with zero porosity. The more complex the part or structure, the greater the likelihood of additional manufacturing flaws, such as wrinkling or bridging. The assembly process – bonding or fastening – is another area where flaws can be added. Even handling and storage, if not done properly, can introduce manufacturing damage. Stress concentrations from flaws and damage can cause the failure of composite structures.

Not all flaws and damages are defects. The term "defect" implies that the part or assembly is unacceptable. Flaws and damages only truly become a defect once the part or assembly is in a condition where it cannot meet its performance requirements (typically a strength or stiffness requirement). Civil aviation certification authorities will stipulate that a structure must demonstrate that it can meet strength requirements while subjected to a certain level of damage. This topic was discussed in further detail in Chapter 2. In general, regardless of the part being manufactured, there will be some level of flaws and damage that is considered acceptable. Anything above that level is a defect, and must be substantiated, repaired, reworked, or replaced. Rework typically follows instructions that result in the part matching original design. Repair refers to a process of following instructions that result in the part having a new configuration. An example of a rework is if there is fraying around a drilled hole that is re-drilled to a larger size that is still allowed on the original design. Repair would be filling the hole and re-drilling elsewhere.

Composite Structures: Effects of Defects, First Edition.
Rani Elhajjar, Peter Grant and Cindy Ashforth.
© 2019 John Wiley & Sons Ltd. Published 2019 by John Wiley & Sons Ltd.

Defects may be described by what they affect in a composite laminate – matrix, fiber, sandwich, joint, or combination. Defects may also be characterized by the stage of manufacturing when they occurred – in the uncured material, during part manufacture, or in service. The classification by stage provides clues in determining the cause of failure. Some defects may occur in multiple stages, such as with tool drops that cause internal damage which may occur during manufacturing, assembly, or in service. In this chapter, we describe the defects separately but in realty many of the defects do not occur in isolation. For example, fiber waviness may be coupled with porosity and regions of resin richness [1]. As we show in this book, these manufacturing defects can result in significant reductions on the operational lifetime compared to pristine structures. In many situations, the effects of defects assessments are not included at the design stage. If the defects detected are within acceptable limits, their presence must be assumed as part of the design. *Establishing the allowable level of defect is a critical element of the composite design process.* It is not that all defects will necessarily cause failures. However, if defects are in highly stressed regions and interact with the loading in a negative way, it is possible that they can initiate failure of the component. Defects may also degrade stiffness and influence the formation of damage away from the defect site. Defects in composite structure can be reduced by:

1) Careful control of incoming material with detailed specifications and inspection tests during receiving inspection. Standard methods examining the quality of resin and fibers can be used to ensure the quality of incoming materials.
2) Detailed material and process specifications.
3) Manufacturing controls of environment, tooling, and all aspects of the manufacturing process. Critical is establishing procedures and system to record and verify the environmental conditions (humidity, temperature, and chemical contamination). A good reference for environmental controls in a composite fabrication facility is provided by Federal Aviation Administration (FAA) advisory circular 21–26A [2]:

> Unless otherwise validated for the material system in use, the area should be temperature- and humidity-controlled such that the minimum temperature is 65 degrees F [18.3 degrees C] with a corresponding relative humidity not greater than 63 percent and the maximum temperature is 75 degrees F [23.9 degrees C] with a corresponding relative humidity not greater than 46 percent. The temperature and relative humidity values between the minimum and maximum acceptable values listed above should form a straight-line relationship.

While it depends on the industry, the current practice is to set thresholds, or allowable defect levels, for given structures. For example, in the aerospace

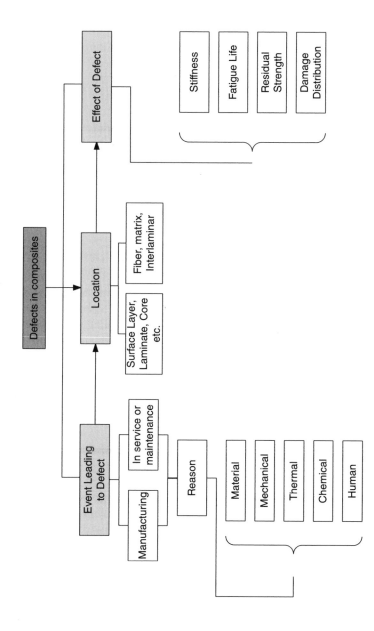

Figure 3.1 Chronology of defects and their effects in a composite structure.

industry porosity is usually not considered a defect if it is less than 2% in volume. It is normally assumed that this level of defect is accounted for in the development of design values. For values higher than this, empirical based data can be used to apply knockdown factors on stiffness and strength for the purpose of determining the effect on structural performance. The complexity of fabrication (storage, forming, consolidation, and curing) can introduce significant variability in the final product. For example, the variability of the fiber volume fractions caused by various defects can significantly influence the mechanical properties of the final product. The design, layup of the material, resin system, curing kinetics, and interaction with the process and tooling can also significantly affect the propensity of the defect to form or to be suppressed.

3.1.1 Differentiating Cosmetic from Structural Defects

Cosmetic defects can be defined as defects that only cause a change in appearance without detrimental effect on the load-carrying capacity of the composite part. It is usually found on the outer exposed surfaces of the composites. Examples include resin ridges, paint chipping, small gouges or scratches, or discoloration. It is usually preferred that such defects be repaired since these defect sites can serve as initiation zones for other types of damage. For example, in sandwich structure a cosmetic surface defect maybe the source of water ingression into the core and that can consequently lead to loss of the load-carrying capacity. Any non-cosmetic defect likely has an impact on structural performance.

Figure 3.1 describes the chronological approach for evaluating the impact of a defect by first considering the event or location dealing with the defect, its location and finally the assessment. Table 3.1 summarizes defects in solid composite laminates, portion of the laminate impacted, and stage of occurrence. Table 3.2 is a similar summary for sandwich structures and Tables 3.3 and 3.4 cover fastened and bonded joints, respectively. Each defect listed in Tables 3.1–3.4 will be defined with its potential sources and effects.

3.2 Defects by Stage of Occurrence

3.2.1 Material Precure Defects

Significant variability of procured material can affect downstream composite processing. Incoming material issues can lead to other laminate or structural-level defects discussed later in this chapter. Therefore, it is critical that target values be established for these quantities with defined allowable deviations. If these values are outside the limits, the cause should be investigated and corrective action taken before using the material again. It is best that these properties are monitored using statistical process control tools to ensure quality of the composite material. Materials should be inspected prior

Table 3.1 Laminate anomalies and associated defects.

Defect	Stage of occurrence		
	Material	**Manufacturing**	**Service**
Bearing damage		✓	✓
Matrix degradation due to aged material	✓		
Blisters	✓	✓	
Matrix degradation due to chemical exposure		✓	✓
Contamination/FOD (includes pre-bond contamination)	✓	✓	
Cracking: corner edge		✓	✓
Crushing		✓	✓
Matrix degradation due to variations in cure cycle (pressure and temperature)		✓	
Composite degradation due to cure cycle disruptions		✓	
Cuts, scratches, and gouges		✓	✓
Delamination		✓	✓
Composite degradation due to excessive temperature exposure		✓	✓
Fiber damage	✓	✓	
Fiber defects (e.g., non-uniform diameter, incorrect sizing)	✓		
Fiber misalignment or wrinkles	✓	✓	
Fiber volume fraction uneven	✓	✓	
Galvanic corrosion			✓
Matrix damage with no fiber breakage from impact		✓	✓
Matrix and fiber damage from impact		✓	✓
Matrix and fiber damage from lightning strikes			✓
Matrix degradation due to anomalous moisture absorption			✓
Matrix cracking or crazing		✓	✓
Matrix degradation due to porosity and voids		✓	
Matrix degradation due to resin mixture error (two-part systems)	✓		
Matrix degradation due to surface contamination (e.g., UV, chemical, erosion, and oxidation)			✓
Matrix degradation due to thermal stresses (i.e., high heat exposure)		✓	✓

(continued)

Table 3.1 (Continued)

Defect	Stage of occurrence		
	Material	Manufacturing	Service
Misdrilled holes		✓	
Mismatched parts		✓	
Missing plies		✓	
Excessive ply drops and g		✓	
Low fiber areal weight	✓	✓	
Warping and thickness variations		✓	
Fiber misalignment or wrinkles	✓	✓	

Table 3.2 Sandwich structure anomalies and associated defects.

Defect	Stage of occurrence		
	Material	Manufacturing	Service
Core defects: core degradation due to water ingression			✓
Core splice: spacing exceeds limits, incomplete core splice		✓	
Edge closeout d: cracked edge members		✓	✓
Edge closeout defect: crushed core at edge member		✓	✓
Facesheet defects: dents in facesheets		✓	✓
Cored: diagonal line of collapsed cells or nested cells	✓	✓	
Facesheet defects: drilled vent holes		✓	
Edge close-out defects: gap between cores or core and edge member, voids in adhesive at edge, incomplete edge seal		✓	
Facesheet defects: pillowing, wrinkling, or orange peel		✓	
Core defects: gaps in machined core/stepped skin		✓	
Defects in adhesive fillets		✓	
Core defects: incorrect ore density	✓		
Core defects: incorrect or variable core thickness	✓		
Core defects: mismatched nodes or corrugations (in high density cores)	✓		
Core defects: over-expanded or blown core	✓	✓	
Facesheet/core disbond		✓	✓
Core defects: misaligned ribbon or unbonded nodes in core cell	✓	✓	

All defects are Sandwich-dominated.

Table 3.3 List of defects in fastened joints and stage of occurrence.

Defect	Stage of occurrence		
	Material	Manufacturing	Service
Bearing damage		✓	✓
Hole delamination/fraying		✓	✓
Hole elongation or out-of-round holes		✓	✓
Fastener seating		✓	
Fastener over-torque		✓	✓
Fastener under-torque		✓	✓
Missing fastener		✓	✓
Porosity near the fastener		✓	
Resin-starved bearing surface		✓	
Insufficient edge margins		✓	
Tilted hole		✓	

Table 3.4 List of defects in bonded joints and stage of occurrence.

Defect	Stage of occurrence		
	Material	Manufacturing	Service
Poor cure due to improper material chemistry: mixing of two-part resins or material past shelf- or out life	✓		
Exothermic reaction in paste adhesive due to mixing of large batches	✓		
Application of paste adhesive outside of pot-life limit	✓		
Amine blush on paste adhesive due to incorrect exposure		✓	
Incorrect bondline thickness, scarf, or overlap length		✓	
Disbonds and zero-volume disbond		✓	✓
Bondline degradation due to contamination from chemicals or moisture or incorrect pressure during processing	✓	✓	✓
Bondline degradation due to incorrect heating procedures	✓	✓	✓

Figure 3.2 Contamination by foreign object on surface of carbon-fiber prepreg.

to cure to reduce the likelihood of defects in the final parts. For example, foreign object debris (FOD) in the uncured material can lead to delaminations, porosity, or wrinkles (Figure 3.2). Visual and instrument-based detection can both be used to ensure that each ply is free from defects before the next one is deposited.

3.2.1.1   Fiber Damage and Defects

It is important to ensure that the incoming prepreg material shows limited variability in its properties. Fibers can be received in defective form with numerous breaks because of manufacturing imperfections. The fibers should be continuous and free from cuts. They should be adequately wetted and there should be no signs of resin richness or starvation. Gaps between fibers or splices are also potential material defects. Fibers in the prepreg should have good alignment and resin and fiber content needs to be uniform and free of any blisters or loose fibers (Figure 3.3).

The fiber areal weight is a function of the density of the fiber, the number of filaments and tows, and the radius of the fiber used. Crimp will also increase the fiber areal weight between 1 and 7% depending on the maximum crimp angle. A low fiber areal weight may indicate resin starvation or low fiber volume content that will impact the composite mechanical properties. Higher crimp angles may also degrade the mechanical properties.

Prepreg should also be free of fuzzballs that are agglomerated broken fibers or chunks of uncured resin that bundle up and can be present in prepreg materials or rolls of the fabric preform. Twisted fiber tows occur when the tows are twisted about the fiber axis and can result in reduced quality of the prepreg. Fibers should also not include wrinkling. This wrinkling refers to the case where the fiber or tow is folded over itself or crosses over another fiber or tow. Wrinkles could also appear as ridges or creases in the prepreg or preform.

The fracture toughness, in-plane, and interlaminar shear strengths have all been shown to increase as the interfacial shear strength between the fiber and the matrix increases [3, 4]. Using contaminated, aged, or incorrect sizings (i.e., fiber surface finishes) can reduce the bonding between the fibers and the resin. Incorrect resin chemistry or mix ratios can affect processing capabilities as well as final mechanical and thermal properties for both laminate and adhesive materials.

The following are additional issues to consider for incoming material defects.

- Material should be within shelf life and out life, and stored at proper environments. Aged prepreg material will likely be difficult to process and is often characterized by a "boardy" texture and increased porosity and voids after cure.
- Correct fibers should be used for the application. In glass fibers, there is little difference in modulus between E-glass or S-glass fibers but the strengths are different. When using carbon fibers, the heat treatment used in the production of the fibers has a significant impact on all the properties. Different grades of carbon fiber can have significant variation in modulus and strength (sometimes up to a factor of 2). The stiffness and strength are usually inversely correlated with heat treatment, so higher stiffness can result in lower strength.
- Perpendicularity: In fabric materials it is important that the warp and fill yarns shall be perpendicular to each other and parallel to the warp and fill directions, respectively. Unidirectional reinforcements should be uniformly aligned along the axis of the material.
- Core material defects include incorrect or variable thickness, which leads to density changes that can result in local instabilities or core failures. Incorrect core density will affect the stiffness of the part, and if incompatible with the facesheets, may lead to disbonds. If different core densities are used on a project, one strategy to maintain core traceability is to use core with the same gage thickness but use core with different cell sizes. This enables simpler checks of the core after the core lamination step since the density can be related to the number of core cells in a specified distance. Core cell disbonds may also affect the shear strength of the core leading to premature failures. Further details of sandwich structure defects are discussed later in this chapter.

3.2.2 Manufacturing Defects

Common manufacturing steps are:

- Part Fabrication/Processing
  - Material handling
  - Laminate ply placement and core assembly (i.e., "layup" or "stacking sequence")
  - Cure or co-bond processing

- Part Assembly
  - Secondary bonding
  - Fastening
- Finish Machining
- Part Transportation/Storage

There are key sources of damage associated with each of the manufacturing steps, summarized in Table 3.5.

Table 3.5 Damage source as function of manufacturing step.

Manufacturing step	Damage source
Material handling	Contamination (including moisture)
	Expired materials
Ply collation	Inclusions (foreign objects)
	High or Low Temperature or High Humidity Environment
	Contamination
	Excessive ply gaps, and overlaps
	Local resin content variations
	Excessive fiber distortion
	Incorrect ply sequence or orientation
	Missing or mislocated plies
	Broken fibers or tows, material cuts
	Inadequate compaction
Cure or co-bonding	Vacuum bag leaks
	Vacuum bag bridging
	Inadequate tooling
	Incorrect cure cycle – ramp rates, dwell times, high or low temperatures or pressures
Secondary bonding	Contaminated bond surfaces or adhesives
	Excessive moisture
	Incorrect bonding process (surface prep, temperature, etc.)
	Incorrect bond gap
Fastening	Poor quality drilling
	Mislocated holes
	Incorrect fastener installation
	Torque setting, fastener grip, and gap pull
Finish machining	Incorrect cut speeds and cutting tools
	Incorrect cooling fluids
Part transportation/storage	Tool drops, large and small
	Edges or surfaces of finished parts damaged
	Contact with tables, vehicles, other parts

Defects such as voids, delaminations, and porosity typically occur during the cure process, and may be the result of poor tooling, insufficient ply consolidation, low autoclave pressure, or loss of vacuum. Poorly designed or installed tools can lead to bridging and lack of proper pressure, leading to areas of porosity and voids as well as wrinkles and delaminations. A common source of delamination is also FOD such as the polymer backing used for prepreg materials.

Some of the latest, tougher materials require ply consolidation (i.e., debulk) prior to the cure cycle. This is due in part to the material viscosity and, depending on the number of plies in the laminate, several compaction cycles may be necessary. In these cases, insufficient ply consolidation, or compaction prior to cure can lead to voids, porosity, and delaminations.

During the cure cycle, any loss of vacuum, autoclave pressure, or temperature can result in voids, porosity, or improper resin wet-out. An improperly cured part may also have low thermal stability in addition to lower mechanical properties.

The primary sources of variability in the curing process, aside from temperature and pressure application, include:

- *Tooling or mold surface finish* where poor surface finish will transfer to the finished product.
- *Tooling materials, density, and spacing of tools* in the autoclave or oven where many tools closer together will act as a heat sink and affect the degree of cure.
- *Part geometry* where the more complex the geometry the more difficult to achieve uniform consolidation and avoid wrinkling or fiber bridging.
- *Stacking sequence symmetry* where non-symmetrical geometry and/or stacking sequence cause part warpage or "springback."
- *Bagging technique and bagging materials* including bleeder materials, cauls, and so on where incomplete material contact against the tool (i.e., bridging) causes non-uniformity in material compaction and resin flow affecting the quality of the finished product.
- *Number of interim debulk cycles and debulk time, temperature and pressure* (vacuum) because insufficient debulking causes thickness and surface finish variability as well as wrinkles in the finished part.
- *Raw material variability* (including batch-to-batch variability), material shelf life, and material out life because material properties are typically time and temperature dependent.
- *Moisture content of materials* being cured or processed because moisture in material affects laminate quality as it turns to steam during cure. The absorbed water may produce dimensional changes (swelling), lower the glass transition temperature of the resin, and reduce the matrix- and matrix/fiber interface-dependent mechanical properties (reference CMH-17, volume 1, Section 2.2.7).

(a)   (b)

Figure 3.3 Examples of incoming prepreg material defects: (a) fuzzy edges or loose fibers and (b) puckers and blisters. Reproduced with permission of Elsevier.

When bonding two or more composite elements or parts, it is essential to ensure that the interface surface of the pre-bonded part is free of any moisture or other contamination. Some OEMs (original equipment manufacturers) utilize peel plies on the interface surface to keep bonding surfaces free from contamination until just prior to secondary bonding or cobonding. Any peel ply used must also be qualified as part of the bonding process. Even after removing the peel ply, some remnant fibers from the peel ply maybe present at the bonding surface.

Macro-voids are easily identified in the fracture surface of a failed bond by the presence of large shiny regions with a smooth glossy surface. Macro-voids are caused by relatively large quantities of gas trapped during the adhesive cure cycle. Common causes for macro-voiding are: incorrect fitment or poor tolerances between the bonding surfaces; inadequate pressurization; or moisture evolution from the adhesive during cure. These kinds of bondline defects are detectable by post-bond NDI (nondestructive inspection). Other bondline defects such as micro-voids, weak bonds, or zero-volume debonds are not readily detectable.

3.2.3 Service Defects

During the operating life of the composite structure, in-service damage may manifest itself in visible or invisible forms. For example, laminate aircraft structures manufactured of carbon or boron fiber-reinforced polymers show that damage can be caused by mechanical impacts [6]. These damages can be caused by ice or runway debris that may impact the structure during takeoffs and landings. Hail, bird strikes, or regular ground maintenance operations may also cause additional damage [7]. Thus composite structures require a quantifiable level of damage tolerance (the ability of the structure to resist a certain amount of damage introduced during manufacturing or service).

Impact damage is of primary concern in composites because they may cause damages that immediately reduce the residual strength of the structure. This is different than the degradation methods for metallic structure, where cracks develop and grow at a predictable rate over a period of time. It is well known that impacts that do not produce visible surface damage can still result in significant damage inside the structure (see Chapters 1 and 2).

Many of the defects discussed in this chapter can result from events occurring during service of a structure and thus must be considered within the overall strategy of design and operation of a composite structure. Primary events leading to damage encountered in the aircraft service environment are listed below. Many of these items are also applicable to sporting or wind energy composites:

- Hailstones
- Runway debris
- Ground vehicles, equipment, and scaffolds
- Tool drops
- Lightning strike
- Bird strikes
- Turbine blade and disk separation
- Heat and fire
- Wear
- Incorrect re-assembly
- Ballistic damage
- Rain erosion
- Ultraviolet exposure
- Hygrothermal cycling
- Oxidative degradation
- Repeated loads
- Fluid ingression
- Chemical exposure
- Improper repair

Hailstones can result in significant damage to aircraft components such as denting fuselage skins, damaging elevators, aileron, or trailing edges. A hailstone can have energy exceeding 33.9 J (300 in.-pounds) and can have sizes more than 5 cm (2 in.) in diameter. Runway debris can be thrown up by landing gear tires and engine blast and can cause serious damage to the undersides of fuselage, wing, and horizontal stabilizer structure. When an aircraft is on the ground it is continually in harm's way from mechanics, ground vehicles, and equipment to other aircraft. There are numerous sources of damage to aircraft structures in the friendly skies from lightning, bird strikes, and the rare engine failure.

All of these stated sources can produce damage: impact damage, scratches/gouges, disbonded elements, burned matrix, and through penetrations. The effects of impact damage can be quite different, depending on the specific design and application. Compression, shear, and tensile strength and stiffness can all be reduced by impact damage. Compressive residual strength of laminated composite material forms is known to depend on the extent of delaminations and fiber failure caused by transverse impacts. Tensile residual strength is affected primarily by fiber failure. Impact damage can also affect the environmental resistance of a composite structural component or the integrity of associated aircraft systems. In addition, those damages may grow during cyclic loading or through other environmental or aging effects.

Structural service damages are repaired when found. However, care must be taken to ensure the repair will return full structural capability. Repairs, like manufacturing, rely upon close process control. Improper repairs may introduce new defects into the structure. Improper repairs may result from using aged repair materials, contamination, or incorrect processing of a repair that escapes post-process inspection (e.g., weak or contaminated bondlines are difficult to detect during post-repair process inspections).

In the aerospace field, the FAA specifies damage tolerance requirements for civilian aircraft. Advisory Circular AC20-107B details five categories of damage and the requirements for certification. These damage categories are substantiated through static and cyclic testing. The general philosophy is that damages that may be present in the structure without repair must be cycled for a full lifetime and then subjected to a static load equal to 1.5 times the maximum load ever expected to be seen in service.[1] Damages that will be found through inspections must be substantiated so that the structure can still support limit load. Other expected/reasonable damages that occur in flight and are known to the pilot must be substantiated for "get home" loads. All other damages are considered anomalous and need not be substantiated for static or fatigue strength (Figure 3.4).

---

1 The maximum load expected in service is called "limit load." "Ultimate load" is 1.5 x limit load.

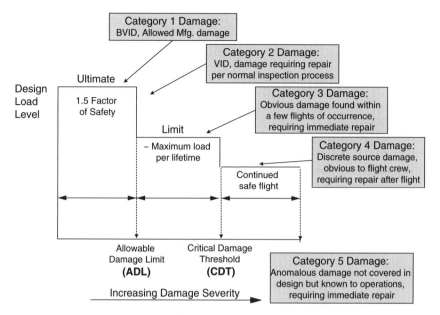

Figure 3.4 FAA damage tolerance requirements [8].

Category 1 damage includes all acceptable manufacturing damage, all damage that could be left unrepaired in service and damage that cannot be detected with the inspection method being used (Table 3.6). This type of damage is often called "BVID" (barely visible impact damage). It is typically limited to 135 J (100 ft lb.) of energy applied with a standard 2.5 cm (1.0 in.) diameter indenter unless probability of detection (POD) studies show detectability at lower energy levels. These damages must be included in fatigue substantiation and shown to hold ultimate load after being cycled for a full lifetime.

Category 2 damage includes likely impact events defined by the actual service environment. These damages are not limited to a standard impact size; that is, can be a blunt or a sharp indenter. These damages would not be found until a dedicated maintenance inspection. Category 2 damage must be cycled for two times the inspection interval and shown to have limit load residual strength.

Category 3 damage is designed to ensure "large damage capability." It encompasses service events that are less likely to happen and, if they do happen, they will be detected within a few cycles, such as during a pilot's walk-around inspection. Note this means that a damage of a certain size may be acceptable on the underside of the wing of a transport aircraft, which is easily visible from the ground, but that same size may be a Category 2 damage if located on the vertical tail, which would not be seen until a dedicated inspection. These damages do not have a standard impact energy or impactor geometry – it is defined by

Table 3.6 Damage categories and structural requirements.

Category	Examples (not inclusive of all damage types)
Category 1: Allowable damage that may go undetected by scheduled or directed field inspection (or allowable mfg. defects)	BVID, scratches, gouges, minor environmental damage, and allowable mfg. defects that retain ultimate load for life
Category 2: Damage detected by scheduled or directed field inspection @ specified intervals (repair scenario)	Visible impact damage (VID) (ranging small to large), deep gouges, mfg. defects/mistakes, major *local* heat, or environmental degradation that retain limit load until found
Category 3: Obvious damage detected within a few flights by operations focal (repair scenario)	Damage obvious to operations in a "walk-around" inspection or due to loss of form/fit/function that must retain limit load twill found by operations
Category 4: Discrete source damage known by pilot to limit flight maneuvers (repair scenario)	Damage in flight from events that are obvious to pilot (rotor burst, bird-strike, lightning, exploding gear tires, severe in-flight hail)
Category 5: Severe damage created by anomalous ground or flight events (repair scenario)	Damage occurring due to rare service events or to an extent beyond that considered in design, which must be reported by operations for immediate action

the service event. These damages do not have to be cycled and therefore may be substantiated during static testing to limit load.

Category 4 damage is defined as discrete source damage known to the pilot. Mitigating actions may be taken and assumed in the strength substantiation (e.g., speed and altitude limitations). The structure need only be substantiated for "get home" loads, typically selected to be approximately 75% of limit load.

Category 5 damage are all other damage sources not considered in Category 1–4. These damages are not included in fatigue and damage tolerance substantiation. The energy magnitude is such that either the damaged structure is obvious, or the energy of the impact is obvious to the parties involved. The structure must be inspected and repaired prior to further flight. Visual inspections and typical SRM (structural repair manual) instructions are not valid for these types of events; dedicated inspections are typically used. The FAA has a policy regarding one type of Category 5 damage, called HEWABI (high energy, wide-area, blunt impacts), described in policy statement PS-ANM-25-20. These types of impacts are specifically related to composite transport aircraft acreage structure where a blunt impact, such as from a baggage loading vehicle, can cause severe structural damage that is not readily apparent from visual inspection of the exterior surface (Figure 3.5).

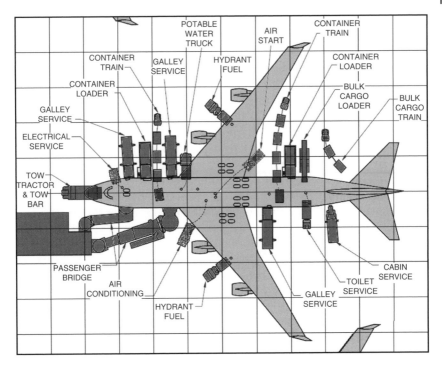

Figure 3.5 Transport airplane ground impact threats.

Damages associated with impact damage include delaminations, disbonds, fiber breakage, matrix cracking, sandwich core crush, and remote reactions/ damage.

- Delaminations typically form at the interface between the layers in the laminate and between face sheets of sandwich structures. Delaminations may form from matrix cracks that grow into the interlaminar layer, from processing non-adhesion, or from low energy impact. Disbonds occur along the bondline between two elements and between the face sheets and the core of sandwich structures.
- Fiber breakage can be critical because composite structures are typically designed to be fiber dominant (i.e., fibers carry most of the loads). Fortunately, fiber failure is typically limited to the zone of impact contact and is constrained by the impact object size and energy. One exception can be a high energy blunt impact that breaks internal structural elements such as stiffeners, ribs, or spars, but leaves the exterior panel laminate intact.

The type and location of damage caused by impact determine the residual compression strength. Broken fibers are the "worst" type of damage due to the loss of load path through the damage site. Delaminations are the second-most critical type of damage because they reduce bending stiffness and buckle locally, leading to failure. Matrix cracks are the least critical type of damage for PMCs (polymer matrix composites). Selecting material and stacking sequence affects how the impact energy is distributed between matrix cracks, delaminations, and fiber breakage.

However, in sporting composite applications, it may not always be possible to produce such damage tolerant behavior after overload events. Overloads are defined as any events deviating from the normal use conditions. Overloads can develop when applying improperly designed patch repairs that can locally result in an overly stiff location that will alter the load redistribution in the part. For example, Trek bicycles does not recommend repair of a carbon-bicycle for exactly this reason [9]. Nonvisible damage can occur from low-velocity impacts such as those from falling tools, equipment, foreign objects, or slow motion collisions. In the case of airplanes and wind turbines, more likely visible damage is caused by high velocity impacts from discrete source events such as part failures from rotating machinery or perhaps ballistic impacts. In a sporting application such as bicycles, the carbon-fiber composites can conceal any damage from an impact or a crash. Bicycle manufacturers typically include a warning to the end customers requiring them to stop using the bicycle after an impact or a crash and to have the bicycle inspected before riding again [9]. The effect of damage can be different depending on the source of damage and its extent. The damage may affect the compression, shear, and tensile strength individually or collectively depending on its nature. Fiber-dominated failures will likely affect the tensile strength and/or modulus whereas damage where the matrix failure dominates the response is more likely to affect the compression and shear properties. The scale of damage relative to the component or structure size should also be considered when assessments are made and load redistribution paths are analyzed. Impact damage can also cause the structure to be compromised thereby allowing moisture or fluids to migrate. Such migration can be detrimental when hybrid structures such as honeycomb composites are used or in the case when the structure acts as a mechanism to retain a particular fluid.

3.3 Defects by Location: Matrix-Dominated Defects

3.3.1 Matrix Degradation Due to Porosity and Voids

Porosity is the presence of small voids, distributed inside a solid laminate. Voids are small pockets of enclosed gas (air, moisture, or volatiles). While the terms "porosity" and "voids" are sometimes used interchangeably, the effects of size can be different in the composite structures. Porosity in this text refers to uniformly distributed pore sizes in the composite structures; whereas voids

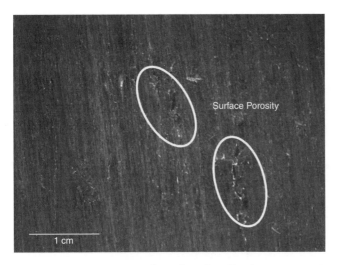

Figure 3.6 Porosity on the surface of a carbon-fiber/epoxy composite.

are used to describe more stratified empty regions. Individually, each void is too small to have a detrimental structural effect; however, collectively, a group of voids can have an adverse effect. Similarly, individual voids are typically not detected by standard nondestructive inspection techniques, but a group of voids may be. The volumetric porosity of primary structural aircraft parts is usually defined in type design, and for autoclave processing it is typically required to be less than 2–3% of the total laminate volume.

Aged/expired material or composite processing can result in porosity or voids present in the final structure (Figure 3.6). They are considered one of the key degradation mechanisms in composite material properties. In liquid molding processing, entrapped voids can be associated with preferential flow channels arising from preform heterogeneities that in turn lead to permeability variations [10]. Porosity refers to defects on the scale of the size of the fibers whereas voids typically refer to larger air inclusions. Voids or porosity can be formed due to air getting entrapped during the mixing of the resin, lack of adequate venting of gasses during cure, low-pressure, thick resin-rich zones in the structure or due to moisture getting entrapped in the material during processing or storage [11, 12]. The moisture can turn into vapor and only show their presence when either vacuum, heat, or both are applied. Vapors can slow down the evacuation and may show as false leaks and if not properly evacuated can cause excessive porosity. Other than water vapor, other vapors can arise from the resin system components.

The voids can also take on different shapes depending on the cause of formation. Smaller randomly distributed voids in the laminate will have a different effect than stratified voids located within interlaminar regions. The resin system

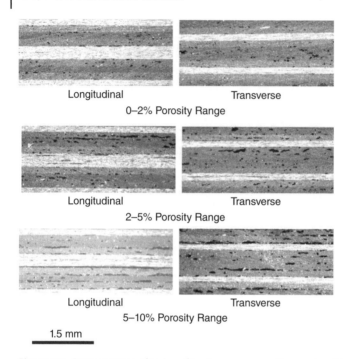

Longitudinal   Transverse
0–2% Porosity Range

Longitudinal   Transverse
2–5% Porosity Range

Longitudinal   Transverse
5–10% Porosity Range

1.5 mm

Figure 3.7 Cross-sections of carbon-fiber/epoxy composite panels with porosity having a quasi-isotropic layup of $[0/45/90/-45/0]_s$ Yang, 2012 [11]. Source: Reprinted with permission from Taylor and Francis.

may affect the shape and size of the voids, which was shown to also to influence the effect of voids on the laminate mechanical properties [13]. Furthermore, sometimes the porosity will manifest itself along the fiber directions as seen in the multidirectional laminate shown in Figure 3.7. It is important to realize that in most cases, local porosity or void content needs to be characterized, rather than global levels. In some cases, the global porosity content in a structure maybe small, but when we examine a smaller area such as near a critical connection detail, the local porosity content is much more of interest when we consider the effect that such levels may have on the local ply compression and shear strengths.

Porosity can have many possible root causes and can occur at various steps in the manufacturing process. The effect of root causes may be compounded, such that two separate steps in a manufacturing process that have in the past produced successful parts in isolation from each other, may lead to porosity when used together. Porosity is also often affected by scaling to large parts; a manufacturing process that was successful at producing small parts and test panels may not be successful when scaled up to complex shapes. Potential root cases are: excess volatiles from raw materials, gaps in layup, insufficient vacuum, insufficient pressure, excess moisture, over aged material, cure cycle deviations, thick laminates, or ultraviolet (UV) exposure after cure (surface porosity).

Moisture absorption in prepreg during the thawing process can also assist in the formation of porosity or voids in the final composite structure. Voids can also be due to off gassing, initial resin air content, release agents, and other reaction products formed as the resin is curing. Lack of sufficient compaction during processing is another source of porosity or voids. This can occur due to many reasons such as bladder leaks, insufficient applied pressure, or processing at high elevations. Moisture is also a known factor contributing to increased porosity due to it resulting in higher vapor pressures. In some matrix materials (e.g., polyimides) where moisture management is an issue, heated platen presses can be used to consolidate at high pressures that can in turn be used to consolidate and minimize porosity formation and growth. In sandwich structure, voids, and delaminations can be caused by: residual moisture on a part due to an improper drying cycle, residual moisture on a part due to too long of a time frame between the drying cycle and the layup and cure, solvent type (especially if slow to evaporate), residual solvent that may have collected in the honeycomb core, residual moisture on the film adhesive and/or prepregs due to exposure to humidity (especially at elevated temperatures, or due to incorrect thawing) [14], or too high a vacuum pressure and/or excessive resin bleed. The higher the vacuum, the more adhesive/resin bleed and lowering of the hydrostatic pressure of the resin that can cause void formation, while excessive vacuum also affects the vapor pressure of any volatiles, thus promoting void growth.

3.3.2 Matrix Degradation Due to Aged Material

It is important to note that thermoset resins and adhesives can cure at any temperature and thermoset prepregs are staged materials that have a finite storage and working out time (Figure 3.8). Solvents or preservatives can be added to resin blends to improve the lifetimes. Longer out times for the materials are needed if the part needs a long time before cure is initiated, as can sometimes occur in large tow placed structures such as large fuselage sections or payload fairings. Two-part resin systems also have finite working life once they are mixed. If the times are exceeded, a thorough review is required to ensure the materials will still perform as required after processing, as degradations in the mechanical properties have to be expected. Unmixed resins (for wet layup or two-part adhesives) may also have reduced storage lives when stored at elevated temperatures.

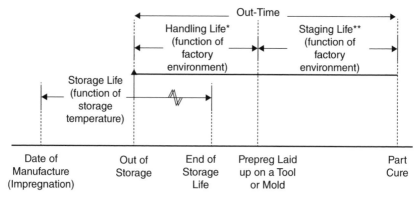

Figure 3.8 Definition of storage life and out life [15] Ward, 2007. Source: Reprinted courtesy of the FAA.

3.3.3 Matrix Degradation Due to Errors in Curing (Pressure and Temperature)

The manufacturing process can influence many of the microstructural features such as the amount of porosity/voids in the final composite structure (including also the extent of crosslinking, fiber volume fraction, and final dimensions). The type of process can also influence the likelihood of these defects with autoclave cures generally producing parts with fewer defects when compared to vacuum assisted curing methods. Autoclave cures provide some of the best quality materials due to the high pressure and good heat circulation that can be applied. In autoclave curing, the high pressure is critical to compact the composite plies and to suppress void formation. In vacuum assisted processing, the pressures are lower and thus voids are more likely to occur. For example, news reports have suggested that the Learjet 85 program, which Bombardier launched in 2007, struggled to perfect the manufacturing of the mostly out-of-autoclave cured carbon-fiber composite design due to porosity issues. According to news reports [16] the parts were built at a facility in Querétaro, Mexico that sits 6000 ft above sea level, which may have impacted the ability to compact the composite structure. A representative curing cycle for a 350°F curing thermoset epoxy typically includes two ramps and two holds (Figure 3.9). The first ramp and hold are used to allow the volatiles to escape as the resin moves from the prepreg phase to achieve its lowest viscosity at that point. In addition to volatiles, air may become entrapped during the layup phase (e.g., at ply drop locations). The second ramp and hold are used for the polymerization process and to allow for a complete crosslinking of the resin. It is essential that the cure cycle not only control the temperature, pressure, and time but also the ramp-up

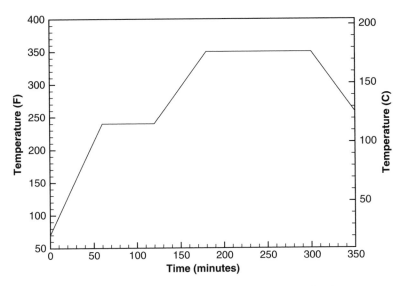

Figure 3.9 Representative thermal curing profile for a commercial carbon/epoxy prepreg.

and cooling down rates. The set time of materials before cure and dwell times are also factors that should be accounted for. Regardless of the curing cycle selected, the key process parameters must be controlled and tailored for the material form developed.

Higher temperature cures are required for many epoxies to achieve the desired properties. The effects of over and undercure are driven by the chemistry of the epoxy. A high cure temperature or too fast of a ramp rate can cause an exothermic reaction. Since epoxies cure through an auto-catalytic process, heat is generated through bond formation and the initial crosslinking process. The extra heat speeds up the curing reaction that also generates more heat. The larger the volume of the structure being cured, the faster the reaction within the epoxy. The exotherms in the epoxy can result in overcure with direct impact on the material properties (e.g., loss in strength, modulus, and hardness). Stepped or ramped cures can help avoid the runaway cures. Unnotched longitudinal and transverse mechanical properties were shown to increase during the cure cycle due to increase in matrix strength and stiffness and development of the fiber/matrix bond [17]. The viscoelastic mechanical response can be strongly dependent on the cure state. At low cure states the creep response is quite significant. As full cure is approached, the material becomes predominantly elastic. Increasing the length of time for a lower temperature cure does not always result in the same degree of crosslinking in the composite as curing at a higher temperature.

Out-of-autoclave composites aim to achieve high quality parts with low porosity (less than 2%). In these materials, the tows are initially dry in the beginning of processing to allow for paths for volatiles to escape but as the process moves to higher temperatures, these gaps are infiltrated. In flat specimens, full tow infiltration was found to occur approximately 50 minutes from the start of the temperature ramp. The air evacuation at the initial stages is critical to achieve low amounts of macro-void content by allowing the air to migrate to the edges [18]. The presence of moisture on the prepreg can also affect the porosity content. One study finds that a relative humidity between 70 and 90% can lead to porosity from 1 to 3% [19]. Further, any anomaly causing changes in the resin viscosity (e.g., materials stored beyond storage or out life [20]) may contribute to greater amounts of porosity in the cured parts.

Applied pressure during cure is critical for properly consolidating the composite. If the pressure is too high, resin migration may occur and dimensional tolerances will be affected. Resin migration can result in lack of wetting or regions with high or low fiber volume fractions. Overcure and undercure of composite materials can lead to significant property differences. Undercure leads to softer matrix properties and weak adhesion between the fibers and the matrix. It can also lead to geometric distortions if the part is removed from the tool while still "soft." Overcure leads to more brittle matrix behavior and a degradation of the resin, which permanently reduces strength properties. A single part may be fully cured in one region, but undercured in another if there was a thermal discrepancy during the manufacturing cycle. The use of epoxy resins in many high-stress structural applications as resin or adhesive requires that a large density of crosslinking occurs in the polymer. The extent of crosslinking density will affect the polymer properties and hence the matrix-dominated properties in the final composite structure. To achieve optimal performance in epoxies, the curing cycle (temperature and time) must be carefully controlled to achieve a high crosslink density. If laminate and consumable materials are exposed to repeated cure cycles or elevated temperatures, they can also begin to degrade. For example, consumable materials, such as the peel ply, which may be on the surface of a laminate for too long prior to bonding can begin to break down and leave a residue on the bonding surface. As the peel ply surfaces are typically bonded at a later step, this can lead to bondline contamination and reduced strength. If the composite part does not go through the required thermal cycle at all locations, some locations may be over or undercured. If the heat transfer conditions change, for example if the air flow velocity within an autoclave is restricted by loading too many parts, then the part may not see the proper thermal cycle and may be undercured in some areas. Parts with multiple cure cycles, such as post-cures or secondary bonding cycles can lead to excessive exposure to heat, which can result in degradation of matrix properties. Use of expired raw materials affects cure behavior, as does improper mixing ratios of two-part liquid resins.

In thermoplastics the material behavior is dominated by the development of crystallinity [21] and development of residual stresses due to mismatches in the coefficient of thermal expansion (CTE) [21, 22]. The effect of cooling rate on the fiber-matrix interface adhesion for a carbon-fiber/semicrystalline polyetheretherketone (PEEK) composite found that the interface bond strength decreased with increasing cooling rate; the tensile strength and elastic modulus of PEEK resin decreased, while the ductility increased with increasing cooling rate through its dominant effect on crystallinity and spherullite size. The properties affected are primarily the matrix-dominated properties such as toughness values [23] but it is important to also check notched strengths since in some cases decreased bond strength can actually increase the notched strength.

3.3.4   Matrix Damage with No Fiber Breakage from Impact

In certain low-velocity impacts, matrix cracking, and delamination are the most common damage mechanisms in laminated fiber-reinforced composites. Research has shown that the composite thickness, dimensions, and impactor will all affect the damage observed with the peak force in an impact event increasing with the thickness of the composite as the contact time decreases [24, 25]. Impact damage should be differentiated from dents where small impressions (e.g., from tooling) occur and where there is no fiber damage. Local reductions in stiffness may occur due to matrix damage.

3.3.5   Matrix Cracking and Crazing

Matrix cracking is defined as intralaminar (local within a single ply) cracks formed in the continuous matrix of the composite. It may form without failure of the reinforcement material. As an example, with an axial tensile load, matrix cracks form parallel to the fibers in a 90° ply. The effect of matrix cracking is reduced structural integrity of the laminate, unless it is localized. It may have a significant effect on resin dominated mechanical properties, such as compressive strength and tensile 90° to the fibers. They could also influence fiber-dominated properties, such as the tensile strength in the fiber direction. Testing is necessary to verify. There may or may not be visible matrix cracks on the surface and NDI techniques may be able to identify regions with matrix cracking. Root causes can be process induced residual stresses due to mismatch of thermal expansion between the resin, fiber, and tooling materials. Cure shrinkage of the resin also contributes to residual stresses. Processing a composite improperly may lead to residual stresses that are larger than the strength of the cured resin, leading to cracking. Other sources may be impact damage, trimming or drilling damage, mishandling, or excessive loading.

Crazing on the other hand, is characterized by the development of a network of fine cracks that occurs in regions of high hydrostatic stress. The direction of the stress can be inferred as perpendicular to the direction of the cracks observed. In thermoplastics, it can lead to increased toughness, but in thermosets it is usually not observed [26]. Thermosets such as epoxy generally do not display crazing due to the high-crosslinking of the epoxy chains that limit the molecular mobility and inhibit the craze formation. However, with the continued development of new thermosets, research indicates that crazing can be made to develop in thermosets containing carbon nanotubes to improve properties [27].

3.3.6 Matrix Degradation Due to Anomalous Moisture Absorption

The effect of moisture on the mechanical properties and the failure behavior of fiber-reinforced polymer composites typically show that moisture decreases the matrix dependent properties [28, 29]. The distinct fall of the matrix- and interface-based values due to moisture can be ascribed to the weakening of bonding between the fiber and matrix and softening of the matrix material [30]. The specimen thickness and layup sequence have been shown to have little effect on the through-the-thickness water absorption behavior of composite laminates [31]. The glass transition temperature of composite laminates is strongly affected and linearly decreased by the quantity of equilibrium water uptake. The following expression has been proposed to estimate the effects of the absorption of general diluetents on the glass transition temperature of epoxies [32].

$$T_g = V_p T_{gp} - (1 - V_p) T_{gd}$$

where, V_p is the volume fraction of the polymer and T_{gp} and T_{gd} are the glass transition of the polymer and diluent, respectively.

Plasticization, swelling, and debonding were identified as the factors affecting the failure mechanisms in composite laminates [33]. In adhesive bonding, pre-bond contamination will affect the Mode-I fracture toughness of carbon-fiber reinforced plastic (CFRP) bonded joints. Release agents and moisture could significantly degrade the Mode-I fracture toughness of joints [34]. Conventional nondestructive testing (NDT) is not always able to detect this type of deterioration at the interface between the contaminated adherend's surface and the adhesive. Thus, it is important that adequate quality control be established for these procedures. Moisture on the uncured materials – particularly core or on prepreg that has been improperly thawed – can also lead to many laminate defects, most notably porosity, and voids (see Section 3.1).

A critical temperature was observed in some studies below which the materials exhibit normal behavior (the rate of absorption of water into epoxy resin increases with increasing temperature) and once the materials saturated with water at temperatures above it, the reversibility of the reverse thermal effect was observed [35]. The effect of temperature on moisture absorption and hygrothermal aging in a commercial bismaleimide (BMI) resin and its composites display a two-stage diffusion behavior, with the first and second stages being diffusion- and relaxation-controlled, respectively [36], with an increase in temperature accelerating the moisture absorption in both the first and second stages. In sandwich construction, moisture ingression can also affect the impact properties. In a study of hybrid carbon-glass fiber-reinforced plastic skins and a syntactic foam core exposure to hygrothermal conditions has shown significant strength reductions for foam specimens and open-edge sandwich panels, compared with reference specimens stored at room temperature [37]. Exposures to non-uniform moisture and temperature can also lead to in-plane deformations and curvature changes of sandwich panels [38].

Moisture absorption during service can be an issue in several structures especially if they also contain large amounts of pores. Sometimes it is not only the matrix but the fiber system is the important parameter to consider. Kevlar fibers are themselves more prone to moisture ingression. Moisture absorption in Kevlar/epoxy laminates has shown a reduction in the glass transition temperature and deterioration in mechanical properties. The strongly hygroscopic Kevlar fiber is frequently a pathway that provides an easy route for moisture ingress [39]. More description of moisture issues in sandwich structure are discussed later in this chapter.

3.3.7 Matrix Degradation Due to UV Radiation or Surface Contamination

If protective coatings on the composite structure are not maintained or are damaged then many of the environmental threats discussed can do damage to composite parts. While UV radiation has little effect on carbon fibers, it can degrade epoxy resins and thus, the integrity of the composite structure can be compromised. UV radiation can cause surface embrittlement of unprotected polymeric composite material. A study on effects of UV and moisture on carbon-fiber/epoxy composite shows that UV radiation and condensation operate in a synergistic manner that leads to extensive matrix erosion, matrix microcracking, fiber debonding, fiber loss, and void formation [40]. The solar radiation incident on the earth surface has UV rays that have a wavelength in the 290–400 nm band. The energy of these UV photons is comparable to the dissociation energies of polymer covalent bonds, which are typically 290–460 kJ mole^{-1} [41].

Composite structures used in aerospace may be exposed to water, anti-icing additive, jet fuel, or hydraulic fluid. Performance of epoxy-based structural adhesives deteriorates considerably when the adhesive is contaminated by hydraulic fluid and anti-icing additive, while the carbon/epoxy weave appears affected by anti-icing additive at a lesser rate [42]. Figure 3.10 shows the diffusion plots for carbon/epoxy lap joints soaked with these fluids and the effect on ultimate tension loads. For epoxy structural adhesives used to bond

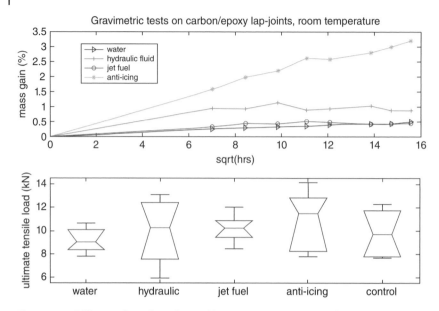

Figure 3.10 Diffusion plots of conditioned lap joints (top), Box plots of ultimate tensile load of control versus conditioned specimens (bottom), Sugita (2010) [42]. Source: Reprinted with permission from Elsevier.

composite and metal parts, the hydraulic fluid exposure results in greater damage at higher temperatures and fuel additive appears to affect the glass transition behavior at all temperatures [43]. The influences of contamination on the mechanical properties of a cured structural carbon-fiber reinforced epoxy composite have been previously investigated [44]. Concentrated mixtures of water and deicing agent, which were prepared in order to simulate the environment in the bottom of fuel tanks, may also reduce joint fracture toughness [45]. However, it is imperative that all fluids (including water) that are relevant beyond this list be considered for the service environment. In core materials, all fluid exposure should be avoided. The effects of fluid exposure should be investigated following methods like those described in CMH-17, Revision G, vol. 1, section 2.3.1.3.

3.3.8 Matrix Degradation Due to High Temperature Exposure

High heat sources can affect composite parts and lead to thermo-oxidative degradation. Examples of high heat sources are: thermal de-icing ducts (typically located in the leading edges of wings), power plants and auxiliary power units (APU), hot air feed ducts, air-conditioning units, and hot air duct failures. If a composite part is heated above its cure temperature, not only are mechanical properties such as stiffness and compression strength

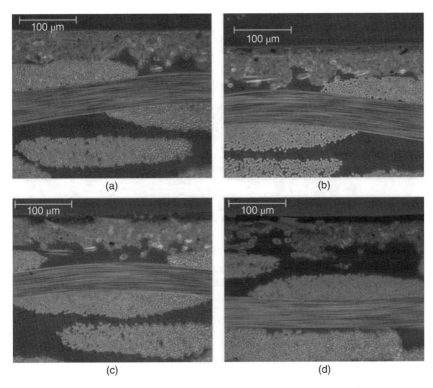

Figure 3.11 Effect of temperature and depth of surface oxidation in a glass fiber/epoxy composite (a) as received. (b) 204°C (400°F) exposure. (c) 232°C (450°F) exposure. (d) 260°C (500°F) exposure [46]. Source: Reproduced with permission from ASM.

compromised, but the epoxy resin may burn resulting in exposure of the fibers, and cracking that can provide moisture or fluid ingression paths. Apart from obvious burn damage, discoloration of the part finish may give an indication of a high temperature exposure. The process typically starts from the surface and proceeds through the thickness of the composite. Figure 3.11 shows the micrograph of degradation in an epoxy woven glass fabric composite with an epoxy surfacing film [46]. In this figure, the original composite (a) is exposed to 204°C (400°F) for a period of time resulting in slight oxidation limited to the surfacing film (b). As the temperature was increased, the oxidation migration into the thickness can be seen in (c) to (d).

Thermal exposure at high temperatures for short periods of time (also called thermal spiking) can also have significant effects on the moisture absorption properties. The effect of moisture absorption as a result of thermal spiking on three composite systems, Narmco Rigidite 5245C, Fibredux 927, and Fibredux 924, has been studied [47]. Figure 3.12 shows how the maximum

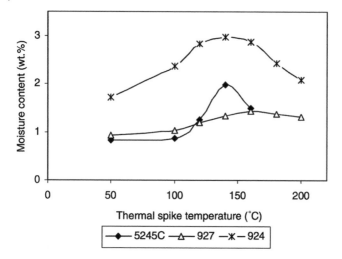

Figure 3.12 Moisture content in the 5245C, 927, and 924 laminate systems, after thermal spiking and conditioning at 96% R.H. for 10,000 hours (5245C/927) and 5100 hours (924) [47]. Source: Reproduced with permission from Elsevier.

moisture enhancement occurring at a spike temperature of 140° and 160°C. No microcracks or voids were observed in the wet laminates that had been spiked indicating that the moisture was absorbed in the polymer network. The thermal spiking also significantly changed the Tg corresponding to significant drops in the storage modulus. The relation between moisture content and the Tg has been seen in other studies involving water and other solvents [48, 49].

3.3.9 Blisters

Blisters are oval shaped features formed on the surface of the composite. They can be as small as sub-millimeter in size and can be found in larger sizes as well. They can occur when a delamination is present near the surface and during cure it is filled with air escaping or from lack of consolidation in the part. Water induced blistering can result from moisture ingression in polymer based composites. The polymer absorbing moisture reduces the glass transition temperature and releases more water from the polymer that causes the polymer to become supersaturated. The moisture can then diffuse into a delamination and by osmotic pressure continue to attract moisture and grow further. A similar mechanism can also occur in the gel-coat layer but those defects are not structure critical. As more water is attracted into the enclosed region the pressure builds inside the blister causing it to grow. Carbon-fiber reinforced polymer composites can form special type of blisters when they are close to metals such as aluminum or steel. In this type of defect, the carbon fibers become cathodes where the meet the metallic parts and salt water may play a factor in accelerating the growth of these blisters [50]. These type of blisters are thought to nucleate at the fiber/matrix interface region [51], which can then fill voids with water or gas and proceed to grow.

3.3.10 Matrix Degradation Due to Resin Mixture Error

Polymer processing typically require the use of a hardener agent to initiate the crosslinking process. Improper mixing can result in significant effects on the matrix-dominated properties of the composite. Improper mixing can result if proper tools such as electric mixers are not used in mixing the fillers and toughening agents that may have settled in the resin during storage. There is also a common mistake that happens when operators confuse the volume or weight ratios. Hardeners typically have a lower density than the resin component. For example, a 5:1 ratio by weight may translate to 3:1 if measured by volume. Unfortunately, the completeness of mixing is difficult to detect, since the colors and viscosities of the hardener and epoxy resin are often quite similar, although some vendors have tried to promote products that can assist in determining this. Mechanical testing of matrix-dominated properties is usually a suggested approach to establish if resin mixing errors have occurred.

3.4 Defects by Location: Fiber-Dominated Process Defects

3.4.1 Fiber Misalignment or Wrinkles

Fiber misalignment or wrinkling is a crease, ridge, or fold in one or more of the laminate plies, where the ply does not rest flat and uniform over the adjacent ply or tool. Ridges on the outside of the cured composite part maybe due to wrinkles or pockets of resin also known as resin ridges. Resin ridges on the exterior of the part, due to imperfections in the tool or creases in a vacuum bag, are not wrinkles and are typically not considered structural defects. True fiber wrinkling can occur at the fiber or ply level. At the fiber level, the wavelengths of the wrinkles are larger and thus can result in significant damage to the composite. At the ply level, out-of-plane waviness are more likely to occur because of the lower bending stiffness in that direction compared to the in-plane directions. Waviness is undesirable curviness of some fibers/plies in an area where all fiber/plies should be aligned. The effects of fiber wrinkling or waviness can simply be a cosmetic imperfection or reduce mechanical properties.

3.4.1.1 In-Plane Waviness
In-plane fiber waviness or distortion is known to occur when changes such as the tension on the rovings (in filament winding or pultrusion) or on the prepreg

tows in automated fiber placement (AFP). Improper handling of unidirectional or fabric materials in hand layup operations can also cause this defect that is also sometimes referred to as a marcel defect. It could also be associated with fiber washing during the resin infusion or transfer molding process or from tool interactions in the part during the cooling process [53]. In filament winding or AFP applications, insufficient tow tension can lead to relaxation and hence in-plane waviness in the structure. Geometry issues such as concave surfaces and corners may also attribute to this defect or if draping of the materials is not done correctly. This type of fiber waviness is also known to occur when there is differential curing in the laminate as documented in an important study on processing parameters [54]. Of the eight parameters investigated – hold temperature, hold time, pressure, length, width, thickness, cooling rate, and tool plate material – three affected the development of fiber waviness: these were length, cooling rate, and tool plate material. For the three relevant parameters investigated, the possible waviness-inducing mechanisms are tool plate/part CTE mismatch, temporal temperature gradients (or cooling rates), and spatial temperature gradients. The tool plate/part CTE mismatch proved to be the most important mechanism driving fiber waviness in plates, although changes in cooling rates also dramatically affected the quantity of waviness. The study confirmed that if the fibers experience axial loads – albeit a small fraction of their Young's modulus – while the matrix is unable to provide some level of transverse fiber support, the fibers will microbuckle resulting in waviness (either in-plane or out-of-plane depending upon the laminate constraint).

3.4.1.2   Out-of-Plane Waviness

In complex geometries such as small radii of curvature or twisting of the structure, if the composite layers are prevented from slipping over one another they can be particularly susceptible to wrinkle formations. Figure 3.13 shows severe cases of waviness in a carbon-fiber bicycle fork illustrating the complex curvature challenges in a small part. Similar factors inducing in-plane waviness can also result in out-of-plane waviness. Draping issues and inadequate tension can lead to these out-of-plane waviness (also sometimes compared to a "carpet ruck"). Friction makes it difficult to harder for large wrinkles to flatten out and may just move in the fabrication process. The plies further away from the tooling surface will experience greater compression and if the resistance is high this can lead to the formation of wrinkles in some of the layers (Figure 3.15). The out-of-plane waviness can occur on the surface as seen in the Figure 3.13 or the case study of the wind turbine failure discussed in Chapter 6, which shows how they can occur deep inside the surface of the laminate. In some structures (e.g., U-shaped parts), wrinkles may form due to shear forces generated as a result of mismatches in the CTE of composite and tool coupled with ply slippage occurring during consolidation [54, 55]. When using multidirectional layups

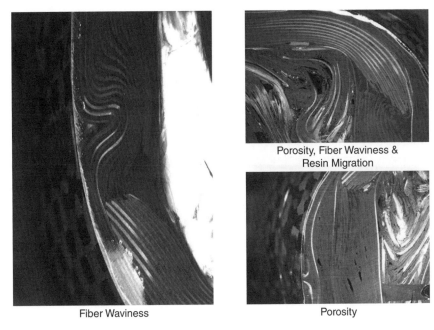

Figure 3.13 Multiple defects in a carbon-fiber epoxy bicycle fork, Sisneros (2012) [52]. Source: Reprinted with permission from Wiley.

out-of-plane and in-plane misalignments can occur. In this study, the authors found that not including the release film results in higher frictional resistance at the tool/composite interface resulting in a lower likelihood of wrinkle formation for the composite material considered. When forming complex contour laminates (e.g., in a composite hat stringer), allowance should be made for slip of the plies otherwise wrinkles may form in the curved sections.

Root causes of wrinkling and waviness, both in- and out-of-plane (Figure 3.14), can be raw material defects or layup errors – especially draping over complex shaped tools. Fibers may be wrinkled or distorted as resin flows into a gap in the layup, with larger gaps and sharper transitions more likely to wrinkle. Improperly placed vacuum bags can cause wrinkles to form under pleats or in corners due to vacuum bag bridging in female corners or pinching in a male corner. During infusion processes, such as resin transfer molding, if the resin feed velocity is too high, the resin viscosity is not low enough, or the fibers are only loosely held, then the fibers may be deformed by the resin flow. On the V-22 FSD (Full Scale Development) the fuselage frame flanges developed significant wrinkles in the flange-web radius [57], probably due to mismatch of the laminate in-plane and out-plane expansion coefficients. These frames were laid up on female tools, which may have contributed to the wrinkle formations.

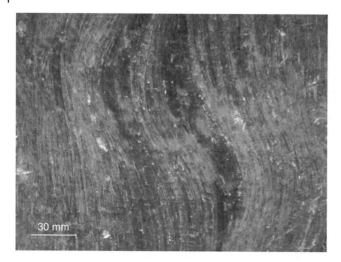

30 mm

Figure 3.14 In-plane waviness on the surface of CFRP/epoxy laminate fabricated from prepreg materials. Note the additional voids also located near the waviness.

3.4.2 Excessive Ply Drops and Gaps

Ply drops and gaps occur in both hand layup and automated manufacturing methods. The AFP technology of fabricating carbon-fiber/epoxy composites is widely used for high performance composite manufacturing. In this method, strips or tows of pre-impregnated slit tape or tows are controlled by an automated deposition robotic head. The head precisely deposits, cuts, clamps, and restarts the composite material on a mandrel driven by a computer-controlled system. The AFP technology allows the customization of the structure by depositing more composite material at areas where higher strength or stiffness are desired. The composite material can also be customized such that it may have different stacking sequences at different locations. The AFP typically deposits the fibers at 0°, ± 45°, and 90° orientations, although other orientations are also possible. The head will typically deposit a number of parallel tows within a single course. During placement buckling of the fibers can occur when steering. When steering on a very smaller radius, the tows will begin to buckle if laid on a flat or a convex surface. On a concave surface, fibers on the inside steering radius of the individual tows adhere to the surface whereas those on the outside are suspended in air leading to a process known as "Venetian blinding" [58]. Ply drops and gaps can also be an issue in hand layup of prepreg or wet layup materials. Simple human error can lead to over- or under-size overlaps. Best design practices avoid butt joints between plies in hand layup due to the potential for low fiber volumes and insufficient load transfer between the plies. Wrinkles can also occur during core splicing in sandwich structure.

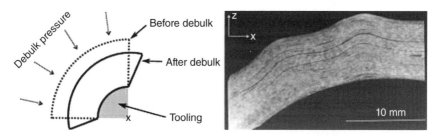

Figure 3.15 Formation of out-of-plane wrinkles during forming on a tight radius due to bookend effect [56].

During the manufacturing, if plies overlap when courses are placed, thicker regions can develop resulting in a higher localized stiffness. These hard zones can be reason of concern since they can attract more loads to those areas. However, if there is a gap between the plies, resin can fill these areas and can contribute to unsupported plies in further manufacturing steps. Ply drops should be as close to the mid-plane of the laminate as possible to reduce the disruption of symmetry. The distance between the successive ply drops should be approximately 10–15 times the dropped height to avoid additional stress concentrations (Figure 3.16). The unsupported plies can lead to localized wrinkles or insufficient load transfer between plies. In addition, the process of cutting plies and customization can result in the formation of wrinkles. For example, consider the case where discontinuous plies can serve as waviness trigger zones when new layers are deposited over them as seen in Figure 3.16. The ply gaps or

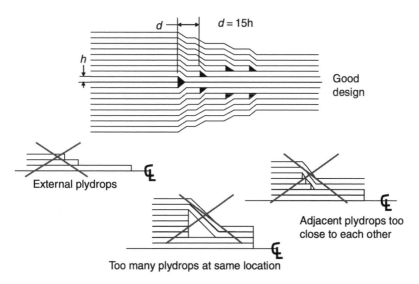

Figure 3.16 Ply drop guidelines to reduce formation of defects [59]. Source: Reproduced with permission from John Wiley & Sons, Ltd.

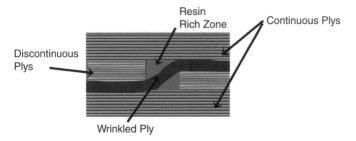

Figure 3.17 Ply gaps and formation of waviness defects.

drops can contribute to the degradation of the composite by allowing the crack to start from the resin-rich zone and propagating along the ply interfaces. The cracks may be arrested if the driving fracture forces are not enough for the crack to propagate. The local reduction of thickness in the case of ply drops can also increase the stress concentration near the discontinuity. Failures in ply drop cases follow similar behaviors observed in other wrinkles under compression where the fiber waviness and ply terminations lead to initiation and growth of delaminations followed by kinking of the wavy fibers followed by buckling and compression failure [60] (Figure 3.17).

3.4.3 Fiber Damage

Fiber damage can occur as an incoming material defect but can also form during manufacturing or service. Improper fiber handling can lead to damage during the processing stages, whereas during service impact damage, gouges, or scratches can result in fibers being damaged. Many composite properties, such as the tensile stiffness and strength, are dependent on the fiber integrity and significant damage to the fibers can compromise the composite structure through the degradation of fiber-dominated mechanical properties.

3.5 Defects by Location: Sandwich Composite Defects

Sandwich composites using metallic or polymeric honeycomb or foam, syntactics, or wood cores are used widely within the composites industry because of their efficient approach to carrying flexural loads. The high stiffness is achieved by increasing the moment of inertia in the part by separating the composite facesheets from the neutral axis of the cross-section. The facesheets are loaded in tension or compression, whereas the core is primarily required for carrying the shear loads. Commonly, the bond between the facesheets and the core is made by using layer(s) of adhesive film. Honeycomb core is preferred for many applications because of its efficient construction mimicking a beehive. In general, it provides better performance than foam cores, but it is usually more expensive and more difficult to process. Adhesive bond failure is a critical failure mode in sandwich structures as it can be associated with three failures modes that depend on adhesive. Facesheet debonding, unbonded nodes, or adhesive fillet failures are all dependent on having good bonding. Degradation of the bond strength can lead to premature failure in the sandwich structure by altering the failure mode from a core failure in shear (typically preferred) to one driven by one of these other failure modes. Identifying corrosion resistant materials and developing moisture ingression resistant closeouts are critical for long-term durability and damage tolerance. These issues are discussed in the following sections.

3.5.1 Core Defect: Over-Expanded or Blown Core

A blown core can occur during the manufacturing of a composite structure if a large pressure differential occurs between the inside of the core and the outside environment. This is more of a risk in autoclave processing when a bag leak occurs [14]. A blown core can lead to facesheet/core disbonds if moisture is present leading to steam pressure builds up in the core at elevated temperatures. Pressure changes can also be caused by changes from ground-air-ground (GAG) cycles or from conditions such as gust loads applied to the structure [61]. The energy release rate ahead of the crack front can be the main driver of the delamination process between the facesheet and the core and maybe influenced by the curvature of the part.

3.5.2 Core Defect: Core Crushing or Movement

In the case of honeycomb composites, the core that is weak in the perpendicular direction can crush or move during processing if not properly designed (Figure 3.18). The mechanisms controlling the core crush sequence are complex and can be affected by the stiffness of the core/facesheets, tooling, and the prepreg frictional resistance [62, 63]. The level of the vacuum pressure and the internal pressure within the core has also been shown to influence the core-crushing phenomenon [64]. Core crushing and disbond can also occur when the structure is subjected to impact. The resultant damage from this mode of loading can influence the mechanical properties associated with compression after impact [65]. The movement of the core during processing can lead to large voids and cracks in the sandwich structure's composite facesheets. The cores may experience movement before or after the resin has gelled. If the cores move right after gelation but before complete cure, the core movement can cause ply separation and resin cracking because the resin has not attained its complete strength at that point.

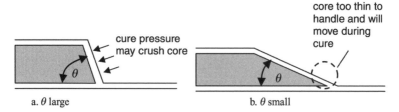

Figure 3.18 Effect of ramp angle on core crush behavior, Kassapoglou, 2013 [59]. Source: Reproduced with permission from John Wiley & Sons, Ltd.

3.5.3 Core Defect: Core-Splice Spacing Exceeding Limits

When large sections of core are desired, sometimes these can be spliced and joined together with adhesives. However, this can result in parts with lower strengths if there are excessive gaps. Sometimes cores with different densities are also needed in the same structure in places where fasteners, backing plates, stiffeners, or other mounting hardware maybe incorporated. The production specifications should specify a maximum allowed gap. Curing of the core-splice adhesive in the case of excessive gap size can cause out-of-plane fiber wrinkling on the surface plies. Typically, the gaps are controlled at less than one cell diameter (or 0.25 in.) of spacing and are filled with adhesive. Disbonds in the core-to-core bond area have been historically limited to five percent of the core-to-core bond area [66]. Main properties affected are the facesheet compression strength and through-thickness shear strength of the assembly. These can be evaluated by test data.

Foam or sandwich core construction can be vulnerable to resin-rich areas if foam cores are undersized or there are improper splices. Resin can pool in imperfection locations and result in parts that are heavy or have areas of low consolidation, as the supporting structure is not rigid. The viscosity of the resin can affect the ability for the resin to pool and cause undesirable effects. A very low-viscosity resin will more quickly pool in in the resin-rich vulnerable areas. The resin must be chosen to provide low viscosity during the resin flow period of the process but maintain sufficient viscosity to remain in place during the manufacturing process. Uneven composite thickness or dry spots can also occur and the increased thickness in some areas can result in overheating during cure. In honeycomb construction, if inadequate support is provided to the facesheets, similar pooling of resin may occur. In either case, if the resin pocket is confirmed to have occurred, an assessment of the fiber wrinkling is required.

3.5.4 Core Defect: Incorrect or Variable Core Thickness

Incorrect thickness can significantly affect the stiffness of the structure as the moment of inertia of the sandwich structure is to a large extent determined by the core thickness. The effect on stiffness and in turn strength is significant and even small reductions in core thickness can lead to significant decrease in bending stiffness.

3.5.5 Core Defect: Core Degradation Due to Core Defect – Water Entrapment in Core

Within aviation, one significant service issue related to moisture is with sandwich structure. Moisture can be absorbed into the core on the ground at elevated temperature and humidity conditions. That moisture then condenses and freezes at flight altitudes. The freeze–thaw cycle can lead to core/facesheet disbonds and/or a change in the mass of a structure. If the sandwich structure is a control surface, this could even lead to aerodynamic instability of the aircraft. This was a contributing factor on the Airbus A310 that lost its rudder due to flutter [67].

3.5.6 Core Defect: Incorrect Core Density

The density of the sandwich structure core is of critical importance to the mechanical properties. Variations in the density during production or collapsed cores may increase the density. When foam composites are considered, sometimes the three-dimensional foam expansion processes can result in variability in the foam density. The foam density is directly linked to many of the important core mechanical properties. For example, the powder blowing method where a foamable powder is inserted into the honeycomb and later caused to foam with the application of heat can result in variability of density from one honeycomb cell to the next. Collapsed cells or potted repairs can cause local increases in core density, but typically these have not been found to be very significant.

3.5.7 Core Defect: Misaligned Nodes or Unbonded Nodes in Core Cell

Misaligned nodes in the core can cause degradation in the core stiffness properties. Degradation of the adhesive bonds between the ribbons can lead to degradation bonds between the honeycomb nodes and consequently result in reduction of the shear properties of the core. Freeze thaw cycling has been shown as one mechanism that can cause the core cells to disbond when water is present in the cells of the honeycomb [68]. The disbonds between the cell walls occur at the cell wall junctions in the corrugated sheets bonded together during the honeycomb structure fabrication process. However, other studies have shown that proper design can show freeze–thaw to not be a significant issue if moisture ingression is controlled [69]. In the case of aluminum cores

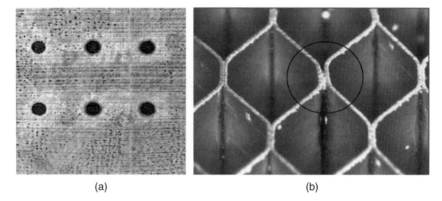

(a)   (b)

Figure 3.19 (a) Ultrasonic image showing core node disbonds (dark circles are 5 cm diameter potted core). (b) Cross-section showing a core node disbond in an aluminum core. [70].

used in sandwich construction, node disbonds can occur during the autoclave cool down due to the mismatch in the CTE between the carbon-fiber/epoxy face sheet and aluminum core [70]. Degradation of the node bonds by affecting the shear modulus of the core can cause premature failure due to shear buckling, crimpling of the core, or even reduce the facesheet wrinkling resistance of the sandwich panel (Figure 3.19).

3.5.8   Core Defect: Mismatched Nodes or Corrugations

Mismatched nodes or corrugations in the core may occur especially in high density cores. Mismatched nodes can lead to unbonded nodes and reduced core shear properties.

3.5.9   Core Defect: Corrosion

Moisture in the core during manufacturing or from service can migrate to the bondline and cause corrosion at the metallic interface if metallic facesheets are used. When metallic cores are used, similar concerns must also be addressed. In metallic honeycomb cores, serious in-service conditions can occur by water ingression into the core causing corrosion of the thin-section core cells. Water ingression can also lead to core cell disbond or delamination in non-metallic cores driven by development of high steam pressures inside the core (e.g., during a repair).

3.5.10   Facesheet Defect: Pillowing, Wrinkling, or Orange Peel

Facesheet wrinkling is typically a catastrophic failure mode in sandwich structures because little or no load-carrying capacity remains after this failure mode

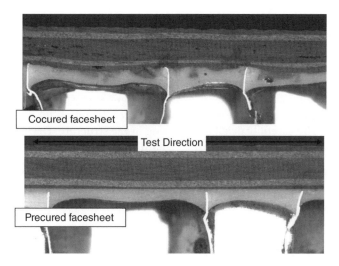

Figure 3.20 Cross-sections of face sheet [+45, 0, −45, 90]$_S$ from co-cured (top) and pre-cured (bottom) sandwich specimens [72]. Source: Image courtesy of NASA.

occurs. It occurs in light core sandwich structures with thin facesheets. In the simplest analysis, it can be considered as a localized short wavelength-buckling phenomenon. During cocuring of prepreg and core, pillowing, or wrinkling of plies into the core may occur. Figure 3.20 shows the difference between pre-cured and cocured panels illustrating the pillowing defect that can occur. The amount of pillowing is going to be a function of the core cell size and other factors such as the facesheet's chemical and physical properties. Initial imperfections in the facesheets can trigger premature failure of the sandwich either by causing a facesheet-to-core flatwise failure (symmetrical imperfections) or a core shear failure by causing an increase in the core shear stresses (antisymmetrical imperfections) [71]. Adhesive degradation or failure can trigger this failure mode by facesheet debonding or adhesive fillet failures. Unsupported facesheet is then more likely to experience this premature failure mode.

3.5.11 Facesheet Defect: Dents in Facesheet

Sharp and abrupt dents on the surface may indicate damage to the facesheet and core. Gradual depressions maybe acceptable if not accompanied by damage in the core and facesheets. Creases, sharp edges, or wrinkles associated with dents are not acceptable without further investigation.

3.5.12 Facesheet/Core Disbond

Disbonds can occur during manufacturing (e.g., moisture in core, failed adhesive fillets, etc.) or during service due to impacts. Disbonds can trigger many failure modes in sandwich structure such as disbond growth and reduction in stiffness of the sandwich structure. Facesheet debonding can also result in premature instabilities such as facesheet wrinkling in the composite structure.

Moisture in the core during manufacturing can migrate to the bondline and cause porosity in the bondline or in the facesheets. Fastener holes, poorly performed prior repairs, or poor sealing may also be water ingression paths. It can also cause blistering, skin disbonds, or result in blown cores. Materials used in manufacturing should be carefully dried before using in the structure. Drying of honeycomb assemblies is difficult because it is hard to determine when the moisture has been completely removed and presence of moisture can also result in failures in any repair procedures [73]. Special attention should be addressed for moisture during repair applications. Moisture absorbed into the adhesive and the elevated temperature used during repair procedures may contribute to a lower adhesive strength during a repair. The high temperature may also increase the pressure inside the core contributing to facesheet separation. However, this can be alleviated by using lower temperature cure adhesive.

3.5.13   Defects in Adhesive Fillets

The mechanism of core and skin delamination is dependent on proper adhesive fillet deformation. Improper adhesive fillets can reduce the delamination resistance of the sandwich composite [74]. Adhesive fillets play an important role in absorbing energy due to fracture in sandwich construction. Honeycombs with poor fillet formation can experience multiple failure modes, especially if the bond strength of the prepreg/film adhesive is not adequate. Improper fillets can include one-sided fillets, short core with no fillets, uneven fillets, and small or underformed fillets. Small fillets can be formed due to poor resin flow or from adhesive that runs down the wall if there is excessive flow [73]. Figure 3.21 shows these multiple failures where the film adhesive separates from the prepreg and areas where the honeycomb cell wall has separated from the adhesive because of improper fillet formation. Adhesive fillet failures can lead to facesheet debonding and can result in premature instabilities in the composite structure.

3.5.14   Edge-Closeout Defects

Closeouts in a sandwich structure must be carefully sealed so that water ingression does not occur. In failure reports in the literature, have shown that panel edges were often found to be the source of most defects [75]. Concepts for edge closeouts shown in Figure 3.22 illustrate both the historical methods used (A–C) and more robust concepts (D, E) that can resist moisture ingression. In concepts A–C, poor tolerances, material issues, and thermal expansion differences in the edge details can all contribute to creating a leak path [69]. The

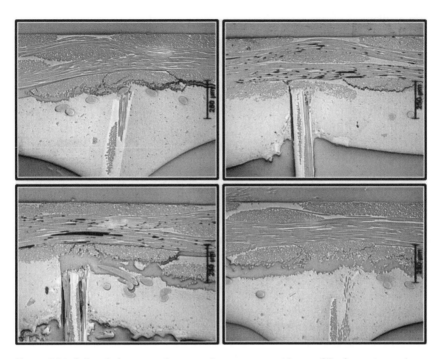

Figure 3.21 Failure in honeycomb composite structures with poor fillet formation and inadequate bond strength between prepreg and film adhesive. Source: Permission ASM.

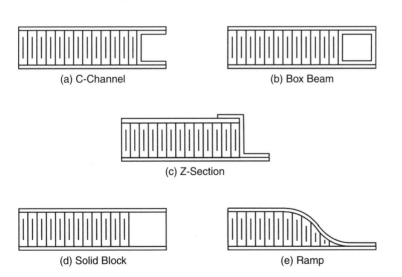

(a) C-Channel

(b) Box Beam

(c) Z-Section

(d) Solid Block

(e) Ramp

Figure 3.22 Options for honeycomb panel edge closeouts, Fogarty, 2010. Source: Reproduced with permission from Springer.

ramp concept (E) is most resistant to moisture but requires contending at the design stage with the forces produced from the joggle and movement of the panel's neutral axis at the edge. The pressure at the edge closeouts that is typically used may not be sufficient to successfully cure the closeout details. Pressure at ramped edges can cause the core to crush, which can be minimized by using foams or other fillers in these regions. However, these can also result in increased weight that maybe undesirable. Other methods include the use of film adhesives to increase the stability of the core at the closeout regions. Fillers added to the edges to facilitate the process can increase the rigidity but can also impact the quality and the compaction process in different ways. If the pressure is not uniform at the end, the laminates in the sandwich construction may experience resin migration issues. Limiting the cocuring pressure can reduce the likelihood of core crushing but can increase the likelihood of porosity formation in the laminates due to improper consolidation as well as the possibility of a poor bond between the facesheet and the core structure. In addition, curvature of the structure in the regions containing the reinforced core can cause it to crack prematurely.

3.6 Defects by Location: Mixed-Mode Fiber and Matrix Defects

3.6.1 Impact Damage with Fiber Breakage

Impacts can lead to damage states where there is a large amount of fiber damage, matrix cracking, and delaminations. This type of damage will impact both the resin- and fiber-dominated properties. Typically, the impacts causing this damage will be at higher speeds and the damage will be visibly detected (as opposed to impact damage without fiber breakage). In addition, any type of grinding operations such as removal of paint or even scheduled maintenances may induce scratches in the composite structure and cause damage to the fibers. In commercial aircraft, a large probability for impact damage can occur when the aircraft is parked at the gate. The damage can be caused during deicing, cargo operations, fueling, or interactions with other ground vehicles. These situations can result in scratch or "gouge" type damage induced on the composite components. If the scratches are only in the matrix rich area on the surface and the fibers are intact then the strength reductions are not as significant as would have occurred if the load-bearing plies had severed through fiber breakage. For the purpose of structural analysis, it is important to note that test results indicate that impact damage produces a greater reduction in capabilities than assuming equivalent hole sizes for encompassing the damage [76]. Strength values for "strength after impact" can be calculated by following ASTM D7137.

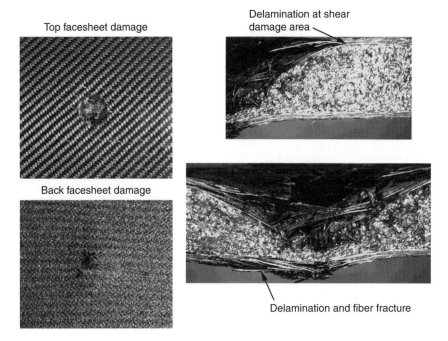

Top facesheet damage

Delamination at shear damage area

Back facesheet damage

Delamination and fiber fracture

Figure 3.23 Multiple damage modes in sandwich composite with carbon-fiber epoxy facesheets and a foam core subject to 50 J of impact energy (Energy of 50 J) [77].

Figure 3.23 shows how visible damage can hide significant internal damage in foam core materials. The core material, cell size, and/or density will all have an impact on the extent of damage tolerance shown by a composite material.

3.6.2   Bearing Damage

Bearing damage refers to damage resulting from bearing loads in a composite structure loaded parallel to the plan of the structure. In the case of a hole, the bearing loads refer to the loads perpendicular to the axis of the hole. Bearing damage is a mixed-mode failure scenario that encompasses fiber micro-buckling, matrix cracking, delamination, and out-of-plane shear failures. Figure 3.24 shows bearing damage in a carbon fiber composite lug. Note the multiple failure modes present. It can occur due to improper joint design such as not performing the proper load share analysis or from fastener overload. The bearing strength capacity is usually established from testing and is dependent on many factors, such as configuration (single or double shear), type of fasteners, and the loading conditions.

When considering repaired structure, where the bearing load at the fastener must be transferred to a shear load at the scarf joints it is important to consider

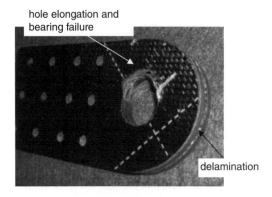

hole elongation and bearing failure

delamination

Figure 3.24 Bearing, hole elongation, and delamination failures in a composite lug, Kassapoglou, 2013 [59]. Source: Reproduced with permission from John Wiley & Sons.

the interaction of both the bearing load capability and stress transfer using shear. Continual detachment and re-attachment of parts can cause damage to fastener holes and adjacent fastened areas. Hole elongation can occur due to repeated load cycling in service and is usually accounted for in the bearing and bolted joint design values. Beyond these accounted for conditions, damage to fastener holes can also happen during maintenance when removing or replacing screws or quick-release fasteners. When this type of damage is localized, it will have limited effect on the performance of multi-fastener bolted joints.

3.6.3 Edge Cracking and Crushing

Corner edge cracking and crushing is a mixed-mode failure that can encompass a wide variety of failure modes including fiber fracture, fiber kinking, matrix cracking, and interlaminar shear failures. Crushing failure modes are primarily governed by interactions between splaying and fragmentation [78]. An impact damage scenario for edge damage follows a similar behavior after development of initial failures from the impact. During an edge impact, initial failures from resin failure, kink-bands, and fiber fragmentation leading to the formation of a wedge made from this debris. These wedge forming failures occur from the initial impact, with fibers broken from compression, shear, and bending [79]. In the splaying failure mode that follows, this wedge forces the delaminated lamina to splay on both sides of sharp cracks. Edge impact damage may be covered by BVID criteria in some cases. In addition, the damage limits on this type of defect can be defined in terms of the dent length and depth values. High resolution computed tomography scans show the damage evolution in a chamfered composite through two crushing distances (Figure 3.25) [80]. At each crush distance, the specimen was removed and examined revealing first the crushing of the 45° outside chamfer providing the debris wedge that results in the significant damage and splaying of the lamina bundles as the crushing continues.

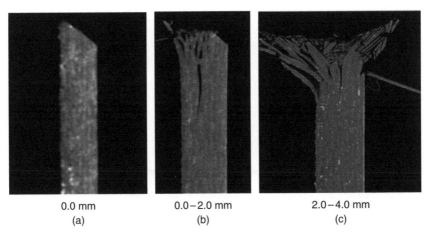

| 0.0 mm | 0.0−2.0 mm | 2.0−4.0 mm |
| (a) | (b) | (c) |

Figure 3.25 High resolution CT scan images of a carbon-fiber based composite specimen (a) before and after crushing distances of (b) 0–2 mm and (c) 2–4 mm. Source: Johnson, 2010 (Reproduced with permission from Taylor and Francis).

3.6.4 Cuts, Scratches, and Gouges

In a symmetric laminate, the mechanics of failure are changed drastically in the presence of deep scratches or gouges that disrupt the load-bearing plies [81]. An optimized aerospace structure usually employs very thin cross-sections in large areas of the body. This type of damage can affect the structural load-bearing capability of the composite structure. Damage analysis to determine the degree of loss of structural integrity is essential for proposing a viable repair technique. The onset and propagation of delamination are critical for an accurate structural integrity assessment of composite structures.

The loss of symmetry from severing the fibers results in high interlaminar stresses, causing delaminations to form near the scratch tip. Therefore, scratch damage analysis and its effects on the failure probability of composite structures are important for damage tolerance considerations. Simulations on effects of scratches show that deep scratches can result in loss of symmetry in a laminate and in additional crack driving forces (Figure 3.26). When considering repair, it is important to understand not only the strength reduction but also the effects of the repair on the local symmetry of the laminate. The lack of symmetry appears to further drive the crack due to the non-uniform deformation. Wind or water flow around a composite part can cause pitting and erosion on the surface. Sometimes this only affects the surface coating and does not expose the fibers significantly.

3.6.5 Composite Damage from Lightning Strikes

Lightning strikes are a significant damage source that must be mitigated through protection methods like copper and aluminum mesh on the outside

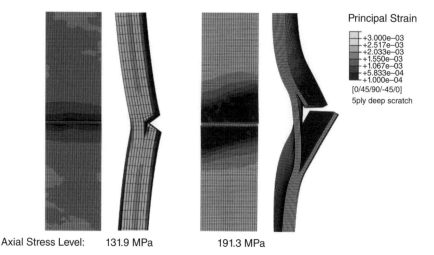

Principal Strain

+3.000e−03
+2.517e−03
+2.033e−03
+1.550e−03
+1.067e−03
+5.833e−04
+1.000e−04
[0/45/90/-45/0]
5ply deep scratch

Axial Stress Level: 131.9 MPa   191.3 MPa

Figure 3.26 Bending and twisting of test specimen at crack initiation and propagation for a specimen with a 5-ply deep scratch, Shams (2013) [82]. Source: Reproduced with permission from Elsevier.

of the structure. Composite structures are less electrically conductive than metals and when hit by lightning, catastrophic damage can occur if they not adequately protected. The high resistivity of composites can lead to highly localized thermal energy buildup that will cause mixed-mode damage including burnout of the resin and fiber pull out and fracture (Figure 3.27).

3.6.6 Misdrilled Holes

Numerous defects can occur when drilling holes during composite assembly. These can include incorrect countersink, oversized, or holes not located properly. Holes that are drilled near the edges are also problematic due to the free-edge effects. Typical rules of thumb in industry range from 2.5 to 3D edge distances. Fiber breakouts and internal damage in the form of a ring of interlaminar failures around the hole can occur if improper drilling procedures are used. These defects can affect the composite allowables.

3.6.7 Mismatched Parts

Sometimes parts do not fit properly and metal, plastic, or liquid shims are used to fill the out of tolerance gaps. It is important to assess the effects of these gaps to make sure no additional loads are introduced into the structure as a result of these gaps.

Figure 3.27 Lightning strike damage in carbon-fiber/BMI composite, Chakravarthi (2011) [83]. Source: Reproduced with permission from John Wiley & Sons.

3.6.8 Incorrect Fiber Orientation or Missing Plies

Composites made from unidirectional fiber prepreg materials are typically designed with specific ply orientations within the thickness. Missing plies or incorrect fiber orientations can lead to significant degradation in the performance. For example, missing a ply in the stack can lead to a loss of symmetry within the laminate cross-section. This loss of symmetry can induce undesirable coupling effects leading to higher stresses than originally factored in the design. Fiber orientation and stacking sequence analysis can be examined using the microscopy methods discussed in the next chapter.

3.6.9 Galvanic Corrosion

Galvanic corrosion can occur when carbon-fiber based composites are in electrical contact with metals. Carbon fibers are electrically conductive. This type of corrosion occurs when anodic materials (e.g., aluminum) are in close contact with cathodic materials (e.g., carbon-composites). Aluminum is particularly vulnerable to galvanic corrosion but so is plain steel. A corrosion production will form on the metal surface that can also induce damage in the composite and cause the development of calcareous deposits. Hydrogen gas formed can result in penetrating any voids within the composite and causing hydrogen or water filled blisters on the composite surface. Galvanic corrosion is minimized through the use of coatings or separating interfaces such as using fiberglass separators. See also Section 3.3.9 for a discussion on galvanic blistering.

3.6.10   Resin Migration and Uneven Fiber Volume Fraction

Resin ridges or pockets can be indicators of more series underlying conditions such as fiber wrinkling. Resin pockets or ridges are defined as areas where resin has accumulated without any reinforcing fibers. In some cases, the ridges can be abraded. Small ridges, generally in practice of less than ~0.02 in. can be considered harmless if there is no fiber distortion. Please refer to Section 3.1, as this defect can be accompanied by fiber distortion that can lead to degradation in the mechanical properties. Imperfection in tooling or manufacturing processes can lead to the formation of regions where resin migrations can occur. Tooling and molds shall be organized such that very sharp corners and edges are avoided, as they can serve as locations where resin can pool (Figure 3.28). Further, avoiding 90° plies at the curved location can help reduce the bridging effects [59]. The absence of reinforcing fibers makes the resin-rich zones especially vulnerable to crack formation and consequently lead to the formation of moisture ingression paths. Resin-rich areas (e.g., in Figure 3.29) may have higher process induced residual stresses and/or deformations due to resin shrinkage. Resin-rich areas also add unnecessary weight. Resin-rich areas can develop in any part of the structure that has a tendency toward a low fiber volume fraction (e.g., ply drops, out-of-plane wrinkles, or in angled parts).

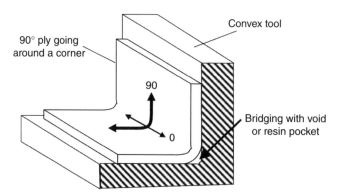

Figure 3.28 Resin pooling at corner of convex tool [59]. Source: Reproduced with permission from publisher John Wiley & Sons.

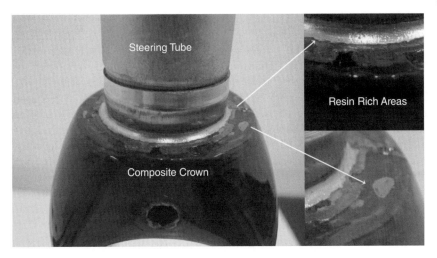

Figure 3.29 Resin buildup in a carbon-fiber/epoxy bicycle component.

Conversely, resin starvation can lead to a lack of the desired uniform resin-rich area on surface and can cause fiber splitting. Proper resin distribution on the surface helps maintain an effective barrier against moisture of the underlying load-bearing composite. Generally, in aircraft structures resin-rich areas will not provide a useful barrier and in fact may lead to excessive moisture absorption, resulting in detrimental freeze–thaw effects.

When incomplete wetting of the fibers occurs, the situation akin to Figure 3.30 can be observed. It is possible to allow a certain amount of resin missing before a repair is required. This may occur when no fibers or only a small amount of fibers are partially exposed. However, in a case where there

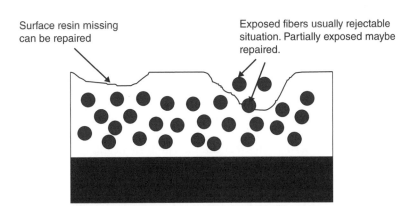

Figure 3.30 Resin starvation at the surface of a composite laminate.

entirely loose fibers on the surface, a repair is usually necessary to restore the bonding between the matrix and fiber and to provide an environmental barrier.

Visual inspection is the primary method to identify resin-rich or resin-starved areas. Some NDI techniques may be able to identify resin-rich or resin-starved areas if there is a local change in the interrogating signal. Destructive techniques, such as micrograph image analysis, may be used to identify and measure resin-rich or starved areas. Root causes are often gaps between layup and tool or between the layup and a caul plate. Pressure gradients, resin bleed off, and insufficient resin added during the wet layup or infusion process can also lead to resin-rich or resin-starved areas. Repair often involves removing the damaged area and installing a repair patch. In some cases, it may be possible to add resin as a sealant over resin-starved areas, in non-structural applications.

3.6.11   Residual Stresses and Dimensional Conformance

Dimensional changes in a composite structure can be the result of residual stresses due to strains from thermal mismatch (i.e., different CTE between plies or between the laminate and the tool); changes in the matrix such as crosslinking or strains from moisture absorption that will also follow the anisotropy of the composite laminate. The stresses can occur at the micro level at the interfaces between fibers and polymers, at the laminate level or the structure level.

At the fiber/matrix level, the different CTE between the fiber and matrix as well as the kinetics of curing can result in different stresses. The application of sizing, fiber architecture, and the volume fraction of the constituents will all affect the level of residual stresses developed at the fiber/matrix level. These stresses may affect the mechanical properties but are unlikely to result in large dimensional changes in the material. These properties are difficult to assess experimentally but can be examined using theoretical micromechanical models that can provide some insights into the stress distributions.

Once the plies are stacked in a laminate, the anisotropy in the CTE between the layers can result in significant development of residual stresses. As the plies in the laminate are trying to expand or contract after processing, the restraint from different layers can results in residual stresses (Figure 3.31). The use of classical lamination plate theory (discussed in Chapter 1) can be used to determine the resulting curvatures. Unsymmetrical or cross-ply laminates can be expected to warp after elevated temperature cure. The determining factors for residual stresses and distortion can be influenced by the stacking sequence, the temperature profile (thick laminates may not have uniform heating), and finally the thickness of each ply.

The tooling used in the fabrication of composites helps ensure that the composite part is uniformly cured. Changes or damage to tooling can result in areas of under- or overcuring in the composite part. The tooling configuration must be controlled to ensure repeatable quality. Having a mismatch in the CTE

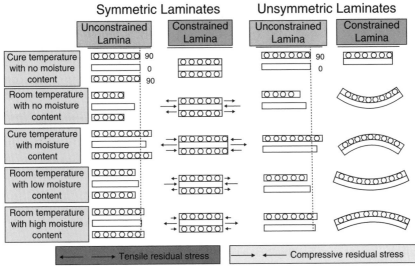

Figure 3.31 Schematic diagram of thermal and moisture effects in symmetric and unsymmetric laminates.

between the tool and the composite part can also result in imposing restraints that lead to residual stresses and dimensional distortions. The effects discussed at the micro and laminate level can also interact with residual stresses produced from the tool-part interaction. The interaction of the part with the tool can also be a source of fiber wrinkling in some parts. During the manufacturing, as the part and the tool heat up, if the expansion of the tool is larger than the CTE of the part, the tool will impose a surface traction force from friction with the part. If the part has a large thickness, the plies closer to the tool will be affected the most. During the low-viscosity stage in the cure cycle, the plies further away from the tool can compensate for the boundary affect. The surface preparation, smoothness, release agent type, and thermal characteristics of the tooling play an important role in determining the amount of residual stresses that can develop and the final warping observed. In cases of unexpected dimensional changes observed, all these factors may have to be investigated.

Hybrid structure, which includes both metallic and composite structure, can be subject to thermal stresses during service, due to thermal mismatch between the materials friction can develop causing large stresses. These stresses can be up to 30% of total load on a part, and are especially significant because they often act in the translaminar direction. For thick composite components, thermal incompatibility between aluminum and carbon/epoxy is a significant issue in joints with multiple fasteners. In aluminum/carbon/epoxy joints with similar stiffnesses, each degree of temperature change can result in internal strain, which becomes significant when considering temperature swings between 50°C (120°F) on the ground and −50°C (−60°F) at altitude. Note that titanium is the most compatible material with carbon/epoxy composites to minimize thermal strains due to its lower coefficient of thermal expansion.

Geometric differences from lack of dimensional conformance are characterized by any loss of dimensional stability of the composite including missing desired tolerances. With composites, due to a variety of processing challenges, the final part shape will often not be the same as the tool or mold shape that it was processed on, leading to challenges in even determining if the part is dimensionally conforming. These defects are important to consider since they can cause issues with assembly to other structures and/or loss in capability in the manufactured part. Warpage or shape distortion can be caused by errors in the ply layup. Unbalanced laminates can be made balanced by compensating plies. In structures with highly curved components a significant degradation can be observed if there is significant mismatch between the CTE of the tooling and composite. Considering unidirectional materials, for example, a large mismatch exists between the fiber (near zero CTE) and transverse material directions. Tooling materials can be selected from aluminum, steel, or composite molding types. Steel and aluminum molds have a higher CTE than carbon-fiber/epoxy and this means that as the part cools down, the tooling will experience larger deformations in shape. Tight dimensional controls may require low-CTE tooling.

Warping in composite parts is minimized when using symmetrical laminates (B matrix = 0) (see Chapter 1). When considering carbon-fiber/epoxy composites, increasing the plies in the off-axis and 90° directions also increases the CTE in the 0 direction. Root causes of dimensional conformance issues can be layup imbalances (asymmetric layups will warp due to non-uniform internal stresses), cure shrinkage of the resin leading to internal stresses which distort the part, tool, and part thermal expansion/contraction, thermal gradients within the part, and tooling dimensional inaccuracies.

3.6.12 Delaminations

A delamination is the separation of two adjacent plies. A disbond is the separation of an intended bondline such as between a core and facesheet or between two structural elements (such as a disbond of a stiffener from a skin – discussed in Section 6.2). Delaminations can be caused by stress concentrations or from free-edge effects associated with the design of the laminate or structural detail. Delaminations similar to matrix defects can affect the compression and shear response and can lead to failure if they grow unchecked.

Root causes of delaminations in manufacturing are often contaminants or foreign object inclusion during ply layup or bonding. Contamination is

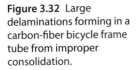

Figure 3.32 Large delaminations forming in a carbon-fiber bicycle frame tube from improper consolidation.

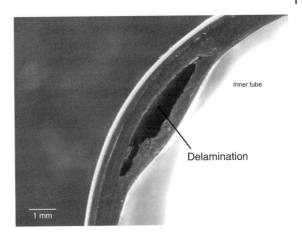

most commonly from excessive moisture in the air during the layup process but can also be caused by free silicone or other aerosols (e.g., if a forklift is operating in the clean room), or from residue of materials that were in contact with the laminate (e.g., peel ply, gloves, lotion on skin, etc.). FOD such as resin or fuzz balls can build up during composite processing – for example, when using automated methods such as AFP. These can form from movement of the AFP head or from cutting operations on the tows. These FOD can become dislodged from the head and find their way on the surface of the structure. A ply-by-ply inspection is necessary to make sure that such FOD does not end up inside the composite structure. If not removed, they will become embedded in the laminate and should be considered as sites for potential delaminations. Visual inspections are typically used after each layer is deposited but incorporating feature recognition and imaging techniques at the AFP head can help in minimizing the risk associated with this defect. Foreign objects can also be in the form of a piece of backing paper left on a ply during layup. It will prevent bonding between adjacent plies. A small foreign object in the ply can initiative a larger delamination, which may form due to an applied load. Improper cure cycles may lead to high internal residual stresses in the laminate. These stresses can exceed the interlaminar strength and lead to delaminations and disbonds. Insufficient pressure application during the cure cycle can also lead to delaminations as seen in Figure 3.32.

Root causes can also be due to improper trimming or drilling parameters; lack of backing materials can lead to push-out delaminations at the exist of the cutting tool or peel up delaminations at the entry of a cutting tool. Impacts on composite parts during handling and storage may lead to fiber fracture and delamination that may or may not be visible on the outer surface. Exposure to excess temperature or thermal cycles may lead to delamination of parts with dissimilar thermal expansion.

3.6.13 Composite Degradation Due to Excessive Temperature and Chemical Exposure

Excessive chemical and temperature exposure can affect both the resin and the fibers and can be compounded by presence of moisture. See also the sections on temperature and chemical effects on the resins and moisture absorption.

3.6.13.1 Fibers

The degradation of fibers can lead to reduced properties. Glass fibers are more susceptible to damage from chemical exposure than carbon fibers. On a study on E-glass fibers exposed to oxalic, hydrochloric, nitric, and sulfuric acids [84], acid corrosion of E-glass fibers was primarily attributed to calcium and aluminum ion depletion. It was found that acid corrosion may generate axial or spiral cracks on the fiber surface, depending on the acid type and concentration. Carbon fiber is more inert and can be degraded when exposed to high temperatures that can occur during a fire. Significant reductions in properties have reported in case of fire in composites [85]. Any type of fiber damage if widespread can lead to a significant reduction in all the fiber-dominated properties.

3.6.13.2 Matrix

The curing of thermosetting polymers results in exothermic reactions. These reactions can result in local temperatures on the composite part to exceed the autoclave (or oven) temperature. The extent of the exotherm can be mitigated if adequate temperature monitoring is used. Nondestructive evaluation (NDE) can be used to determine which parts were overheated or if the overheating was localized. The thermal degradation can also be accompanied by matrix cracking with its direct correlation to impacts on shear and compression properties. Note that most thermosets (e.g., epoxy) used in aerospace char and burn rather than melt. Some thermosets will also degrade under exposure to chemicals such as paint stripper.

3.7 Defects by Location: Fastened and Bonded Joint Defects

3.7.1 Fastened Joints

During the assembly of joints in composite materials, there is a possibility of errors in drilling holes or installing fasteners. In metals, slight dimensional issues are not as problematic since plastic deformation near the fasteners will help distribute the load evenly. However, in composites, a yield point is usually not present and load distribution is not necessary equalized during loading. These may affect the failure loads of bolted joint assemblies for either the composite or the fastener. The defects in the joint installation may affect the following possible composite and fastener failures that may occur independently or occur together.

- Bearing failure of the composite.
- Increased loading on the fastener due to slipping in the joint detail and/or coupled with of bearing failure in the composite laminate.
- Shear tear-out or cleavage failure in the composite.
- Fasteners pull through from the composite or net tension failure for the case of thin sections.
- The fastener may fail under shear or tension loads exceeding the ultimate strengths of the fastener.
- Bolt bending from bending stress induced in the laminate.
- Machining defects associated with holes such as oversize, tilt, or incorrectly drilled and filled holes.
- Improper fastener installation.

3.7.1.1 Bearing Damage
Please refer to Section 3.6.2 for laminate discussion on bearing damage. Most bearing damage occurs during service due to overloads at the fasteners or errors in joint design.

3.7.1.2 Hole Delamination/Fraying
Broken fibers on the exit side of the hole can occur when there are issues with the material, incorrect backing is used, or improper drilling speeds were used. Dull drill bits can also cause hole delaminations or fraying. Delaminations are generally severe when more than 20% of the laminate thickness is impacted on the exit side.

3.7.1.3 Hole Elongation or Out-of-Round Holes
Hole elongation or out-of-round holes can occur when the drill is offset from the correct position by a small amount. Testing indicates little sensitivity for out of roundness for up to 0.1 mm (0.004 in.).

3.7.1.4 Fastener Seating
Excessive countersink (typically greater than 50%) of the laminate thickness can occur during fastener hole preparation. Countersinking greater than this can result in reductions of the bearing capacity because of insufficient hear bearing into the laminate.

3.7.1.5 Fastener Over-Torque
Laminate damage in the forms of matrix damage can occur if fastener is over-torqued. The force in the fastener will also drop at a different rate than if it was installed at the nominal rate.

3.7.1.6 Fastener Under-Torque

Under-torqued fasteners will result in lower joint strength and potential fitting errors if slop develops. Hole wear is also likely to be accelerated if the fastener is under-torqued.

3.7.1.7 Missing Fastener

A missing fastener will alter the load sharing arrangement between the fasteners in a bolted joint assembly and can result in more loads going to the other fasteners. This can result in fastener failure and/or an increase of the local load bearing at the fastener hole.

3.7.1.8 Porosity Near the Fastener

Porosity near the location of the fastener results in severe reduction of the bearing properties and overall joint strength reduction.

3.7.1.9 Resin-Starved Bearing Surface

The resin-starved surface will affect the location of the fastener results in severe reduction of the bearing properties and overall joint strength reduction.

3.7.1.10 Insufficient Edge Margins

The region near the edges of a laminate are zones of high residual interlaminar stresses resulting from the balance of forces in the laminate due to the different Poisson ratios of the different plies. Introducing holes near the edges is not recommended because of the interaction with this high-stress region. Edge distances of at least 2.5–3 times of the hole diameter are typically used.

3.7.1.11 Tilted Hole

Perpendicularity of the countersink is important to control. Deviations from perpendicularity can result in degradation in bolted joint properties.

3.7.2 Bonded Joints

Various defects can be introduced in bonded joint structures, most notably disbonds and weak bonds. Porosity and voids or microcracks can also manifest themselves in the bondline. Figure 3.33 shows the typical defects in bonded joints. A disbond can be classified as either a disbond associated with a volume that can be detected by NDI procedures or it may be a zero-volume disbond, which is difficult to detect with standard methods. A weak bond is defined in FAA AC 20-107B [8] as "a line with mechanical properties lower than expected, but without any possibility to detect that by normal NDI procedures." A poorly prepared bond may appear to be well formed, but if the bond has been contaminated, experienced a poor cure, or suffered inadequate surface preparation, it may not be able to withstand the required loads. Poor bond integrity may lead to premature disbonds and possible structural failures. Bond line thickness can be visually inspected at edge surfaces, but it is very difficult to identify a bond with

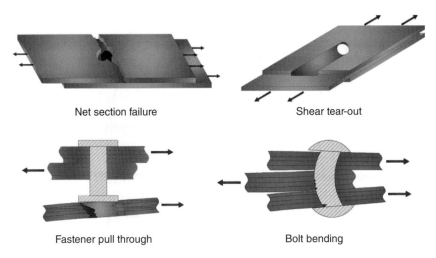

| Net section failure | Shear tear-out |
| Fastener pull through | Bolt bending |

Figure 3.33 Example of joint failures.

poor integrity after the bond has been formed, though some NDI techniques can detect non-zero-volume disbonds.

Root causes of a weak bond can be poor surface preparation, moisture absorption on the bonding surface, FOD, bond line thickness variations, lack of proper contact between the adhesive and substrate during cure, or low quality adhesive materials. Silicone based oils introduced to the bondline can result in the formation of a weak bond [86]. If the affected area is small and repair is allowed, local fasteners may be installed. If the affected area is larger, the part would likely be scrapped.

Traditional ultrasonic methods (e.g., pulse-echo C-scan) that rely on changes of impedance in the medium are not useful in detecting weak bonds since there is no noticeable separation between the adherend and adhesive surfaces. In the case of an effects-of-defects assessment, it is important to identify that (1) the defect does not migrate to the adherend or adhesives, (2) the thickness of the introduced defect is very small to be negligible using typical NDE methods, and (3) the defects are of enough magnitude to cause a degradation in the bond strength (Figure 3.34).

Bonded repairs should be included in the damage substantiation process and, because of limited confidence in making good repairs (e.g., zero-volume disbond), the size should be limited.

According to FAA policy statement in 2014 [88]:

1) The repair process must be performed according to process specifications using approved and qualified materials,
2) The repairs must be substantiated to ultimate load by testing or analysis supported by tests,

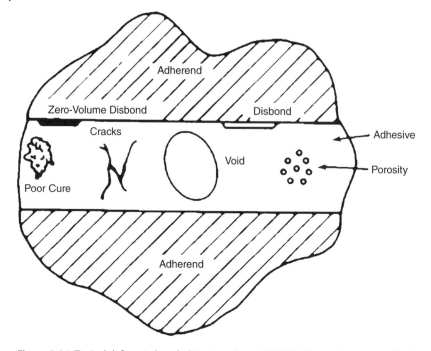

Figure 3.34 Typical defects in bonded joints, Adams 1988 [87]. Source: Reproduced with permission of Elsevier.

3) The data supporting the repair must include inspections that are capable of detecting complete or partial failure of the bond line, and
4) Bonded repairs must demonstrate that if the bondline fails between arresting features, the remaining structure will hold limit load.

This is primarily due to the fact that understrength bonds (i.e., zero-volume disbonds) cannot be detected through nondestructive evaluations.

The adhesive, adherend and the bonding process are typically developed to deliver a given level of performance and changes in any of these materials or processes can lead to a situation where the joint quality is compromised. The following list includes some factors to consider in bonded joints:

- Surface preparation by proper cleaning and abrasion to the surface to prepare for bonding.
- The bond thickness is controlled by adequate compression, improper control of this variable can lead also lead to void formation.
- Storage and pot life of adhesive materials. Also, related to improper mixing of hardener agents.
- Following proper cure cycle.

3.7.2.1 Poor Cure Due to Improper Material Chemistry: Mixing of Two-Part Resins or Material Past Shelf- or Out Life

Incomplete mixing of two-part adhesives can result in non-uniform properties due to variation of the hardener and resin distribution. It can result in lower bond strength due to improper curing or a lack of development of the structural properties of the adhesive by chemical reaction. Resin stored past the shelf- or out life can also lead to degradation in the resin properties and may have issues attaining the proper viscosity necessary for proper wetting.

3.7.2.2 Incorrect Bondline Thickness, Scarf, or Overlap Length

Achieving the exact matching between the two adherends require precise matching of the surfaces especially if pre-cured patches are used. Geometrical or details of the repair that are out of specification can result in a reduced bond strength since the repair is designed to transition a complex stress state between the two adherends.

3.7.2.3 Zero-Volume Disbond or Long-Term Bond Failure Due to Contamination or Incorrect Surface Preparation

Surface contamination is a major issue that can result in a lower bond strength and reduced durability. Epoxy resin can absorb contaminants on the surface such as moisture, hydraulic fluids, release agents, or other contaminants on the surface. The surface needs to be clean of contaminants and prepared for bonding.

A disbond is an area of adhesion failure in the area between the adherends. If the disbond is associated with surfaces intimately in contact, the disbond is referred to as a zero-volume disbond. The disbonds can be caused by contamination or errors in the repair or bonding procedure.

3.7.2.4 Bondline Degradation Due to Moisture or Incorrect Pressure During Processing

Poor pressurization when failures occur in the vacuum bag can result in formation of porosity or voids in the bondline and sometimes in the patch as well. The mechanism is similar to that discussed for laminates due to poor consolidation. Moisture is another cause of porosity or voids in the bondline. It can occur due inadequate drying of the composite or if the environment is excessively humid. The adhesive may also be previously contaminated as well. The presence of the moisture in the core of a sandwich structure can detrimentally impact the bondline durability if not properly designed. The moisture in the sandwich structure can detrimentally impact the bondline strength by causing a disbond to propagate or can result in a blown core due to buildup of vapor pressure inside the sandwich structure.

3.7.2.5   Bondline Degradation Due to Incorrect Heating Procedures

Undercuring of the adhesive will result in lower mechanical properties and a failure to achieved the required bond strength. Also, if the heating is performed too rapidly, the volatiles in the adhesive can get trapped resulting in voids in the bondline. However, slower heating rates can result in poor adhesion resulting in lower bond strengths. Overheating of the adhesive can result in damage to the adhesive resulting in discoloration, charring, or crazing cracks. The overheating can extend to regions beyond the adhesive bondline and can cause delaminations in the composite and damage to the matrix of the composite structure. If the supporting structure contains metallic components, these could also suffer degradation such as in the case sandwich structures containing aluminum honeycomb.

3.8   Future Directions

Further development of tools that can predict manufacturing defects such as porosity, ply waviness, delaminations, fiber wrinkling, and resin starvation/rich areas are likely to result in higher structural reliability and offer the designer the ability to optimize the composite structure. The structural analyst can then use this information to provide input on the effects of such defects on the structural performance. Such simulations will require multi-physics interactions including chemical, thermal, and mechanical effects and must account for the tooling and process environment including the interaction between all components, temperature, pressure, and resin kinetics. Methods for surface preparation, novel heating methods, such as induction heating and conductive patches; health monitoring of bonded composite repairs are areas likely to witness increased innovations to remove or significantly reduce the risk associated with bonded repairs. Ideally, a NDI method will be developed to ensure full bond strength has been achieved.

References

1 Smith, B., R. Grove, and T. Munns, Failure Analysis of Composite Structure Materials. 1986, DTIC Document.
2 FAA, AC 21-26, Quality System for the Manufacture of Composite Structures. 2010, Federal Aviation Administration.
3 Madhukar, M.S. and L.T. Drzal, Fiber-matrix adhesion and its effect on composite mechanical properties: I. Inplane and interlaminar shear behavior of graphite/epoxy composites. Journal of Composite Materials, 1991. 25(8): pp. 932–957.

4 Madhukar, M.S. and L.T. Drzal, Fiber-matrix adhesion and its effect on composite mechanical properties: IV. Mode I and mode II fracture toughness of graphite/epoxy composites. Journal of Composite Materials, 1992. 26(7): pp. 936–968.

5 Cemenska, J., T. Rudberg, and M. Henscheid, Automated in-process inspection system for AFP machines. SAE International Journal of Aerospace, 2015. 8(2015-01-2608): pp. 303–309.

6 Ushakov, A., A. Stewart, I. Mishulin, and A. Pankov, Probabilistic Design of Damage Tolerant Composite Aircraft Structures. 2002, Federal Aviation Administration.

7 Ratwani, M., Effect of Damage on Strength and Durability, 2010, NATO Research and Technology Organisation.

8 FAA, Composite Aircraft Structure, in FAA Advisory Circular AC 20-107B. 2009, Federal Aviation Administration.

9 Trek, Trek Bicycle User Manual. Trek Bicycle Corporation: Waterloo, Wisconsin.

10 Hamidi, Y.K., L. Aktas, and M.C. Altan, Formation of microscopic voids in resin transfer molded composites. Journal of Engineering Materials and Technology, 2004. 126(4): pp. 420–426.

11 Yang, P. and R. El-Hajjar, Porosity Defect Morphology Effects in Carbon Fiber–Epoxy Composites. Polymer-Plastics Technology and Engineering, 2012. 51(11): pp. 1141–1148.

12 Jeong, H., Effects of voids on the mechanical strength and ultrasonic attenuation of laminated composites. Journal of Composite Materials, 1997. 31(3): pp. 276–292.

13 Costa, M.L., S.F.M. De Almeida, and M.C. Rezende, The influence of porosity on the interlaminar shear strength of carbon/epoxy and carbon/bismaleimide fabric laminates. Composites Science and Technology, 2001. 61(14): pp. 2101–2108.

14 Campbell Jr,, F.C., Manufacturing Processes for Advanced Composites. 2003, Elsevier.

15 Ward, S., W. McCarvill, and J. Tomblin, Guidelines and Recommended Criteria for the Development of a Material Specification for Carbon Fiber/Epoxy Fabric Prepregs. 2007, U.S. Department of Transportation, Federal Aviation Administration: Springfield, VA.

16 Ostrower, J. and P. Vieira, Bombardier Halts Learjet 85 Program Amid Weak Demand, in The Wall Street Journal. 2015, New York, NY.

17 White, S. and H. Hahn, Process modeling of composite materials: residual stress development during cure. Part II. Experimental validation. Journal of Composite Materials, 1992. 26(16): pp. 2423–2453.

18 Centea, T. and P. Hubert, Measuring the impregnation of an out-of-autoclave prepreg by micro-CT. Composites Science and Technology, 2011. 71(5): pp. 593–599.

19 Grunenfelder, L. and S. Nutt, Void formation in composite prepregs–effect of dissolved moisture. Composites Science and Technology, 2010. 70(16): pp. 2304–2309.

20 Grunenfelder, L., T. Centea, P. Hubert, and S. Nutt, Effect of room-temperature out-time on tow impregnation in an out-of-autoclave prepreg. Composites Part A: Applied Science and Manufacturing, 2013. 45: pp. 119–126.

21 Gao, S.-L. and J.-K. Kim, Cooling rate influences in carbon fibre/PEEK composites. Part 1. Crystallinity and interface adhesion. Composites Part A: Applied Science and Manufacturing, 2000. 31(6): pp. 517–530.

22 Parlevliet, P.P., H.E. Bersee, and A. Beukers, Residual stresses in thermoplastic composites—a study of the literature—part I: formation of residual stresses. Composites Part A: Applied Science and Manufacturing, 2006. 37(11): pp. 1847–1857.

23 Lee, W.I., M.F. Talbott, G.S. Springer, and L.A. Berglund, Effects of cooling rate on the crystallinity and mechanical properties of thermoplastic composites. Journal of Reinforced Plastics and Composites, 1987. 6(1): pp. 2–12.

24 Aslan, Z., R. Karakuzu, and B. Okutan, The response of laminated composite plates under low-velocity impact loading. Composite Structures, 2003. 59(1): pp. 119–127.

25 Aslan, Z., R. Karakuzu, and O. Sayman, Dynamic characteristics of laminated woven E-glass–epoxy composite plates subjected to Low velocity heavy mass impact. Journal of Composite Materials, 2002. 36(21): pp. 2421–2442.

26 Kinloch, A. and J. Williams, Crack blunting mechanisms in polymers. Journal of Materials Science, 1980. 15(4): pp. 987–996.

27 Zhang, W., I. Srivastava, Y.F. Zhu, et al., Heterogeneity in epoxy nanocomposites initiates crazing: significant improvements in fatigue resistance and toughening. Small, 2009. 5(12): pp. 1403–1407.

28 Ray, B., Temperature effect during humid ageing on interfaces of glass and carbon fibers reinforced epoxy composites. Journal of Colloid and Interface Science, 2006. 298(1): pp. 111–117.

29 Shen, C.-H. and G.S. Springer, Moisture absorption and desorption of composite materials. Journal of Composite Materials, 1976. 10(1): pp. 2–20.

30 Selzer, R. and K. Friedrich, Mechanical properties and failure behaviour of carbon fibre-reinforced polymer composites under the influence of moisture. Composites Part A: Applied Science and Manufacturing, 1997. 28(6): pp. 595–604.

31 Choi, H., K. Ahn, J.-D. Nam, and H. Chun, Hygroscopic aspects of epoxy/carbon fiber composite laminates in aircraft environments. Composites Part A: Applied Science and Manufacturing, 2001. 32(5): pp. 709–720.

32 Turi, E.A., Thermal Characterization of Polymeric Materials. 1997, Academic Press: New York.

33 Boll, D., W. Bascom, and B. Motiee, Moisture absorption by structural epoxy-matrix carbon-fiber composites. Composites Science and Technology, 1985. 24(4): pp. 253–273.

34 Markatos, D., K. Tserpes, E. Rau, et al., Degradation of mode-I fracture toughness of CFRP bonded joints due to release agent and moisture pre-bond contamination. The Journal of Adhesion, 2014. 90(2): pp. 156–173.

35 Zheng, Q. and R. Morgan, Synergistic thermal-moisture damage mechanisms of epoxies and their carbon fiber composites. Journal of Composite Materials, 1993. 27(15): pp. 1465–1478.

36 Bao, L.-R. and A.F. Yee, Effect of temperature on moisture absorption in a bismaleimide resin and its carbon fiber composites. Polymer, 2002. 43(14): pp. 3987–3997.

37 Ishai, O., C. Hiel, and M. Luft, Long-term hygrothermal effects on damage tolerance of hybrid composite sandwich panels. Composites, 1995. 26(1): pp. 47–55.

38 Doxsee, L.E., W.I. Lee, G.S. Springer, and S.S. Chang, Temperature and moisture induced deformations in composite sandwich panels. Journal of Reinforced Plastics and Composites, 1985. 4(4): pp. 326–353.

39 Akay, M., S.K.A. Mun, and A. Stanley, Influence of moisture on the thermal and mechanical properties of autoclaved and oven-cured Kevlar-49/epoxy laminates. Composites Science and Technology, 1997. 57(5): pp. 565–571.

40 Kumar, B.G., R.P. Singh, and T. Nakamura, Degradation of carbon fiber-reinforced epoxy composites by ultraviolet radiation and condensation. Journal of Composite Materials, 2002. 36(24): pp. 2713–2733.

41 Ranby, B.G. and J.F. Rabek, Photodegradation, Photo-Oxidation, and Photostabilization of Polymers: Principles and Applications. 1975, John Wiley & Sons.

42 Sugita, Y., C. Winkelmann, and V. La Saponara, Environmental and chemical degradation of carbon/epoxy lap joints for aerospace applications, and effects on their mechanical performance. Composites Science and Technology, 2010. 70(5): pp. 829–839.

43 Campbell, R.A., B.M. Pickett, V. La Saponara, and D. Dierdorf, Thermal characterization and flammability of structural epoxy adhesive and carbon/epoxy composite with environmental and chemical degradation. Journal of Adhesion Science and Technology, 2012. 26(7): pp. 889–910.

44 Zhang, M. and S. Mason, The effects of contamination on the mechanical properties of carbon fibre reinforced epoxy composite materials. Journal of Composite Materials, 1999. 33(14): pp. 1363–1374.

45 Rider, A. and E. Yeo, The Chemical Resistance of Epoxy Adhesive Joints Exposed to Aviation Fuel and its Additives. 2005, DTIC Document.

46 Hayes, B.S. and L.M. Gammon, Optical Microscopy of Fiber-Reinforced Composites. 2010, ASM International.

47 Hough, J.A., S.K. Karad, and F.R. Jones, The effect of thermal spiking on moisture absorption, mechanical and viscoelastic properties of carbon fibre reinforced epoxy laminates. Composites Science and Technology, 2005. 65(7): pp. 1299–1305.

48 Pérez-Pacheco, E., J. Cauich-Cupul, A. Valadez-González, and P. Herrera-Franco, Effect of moisture absorption on the mechanical behavior of carbon fiber/epoxy matrix composites. Journal of Materials Science, 2013. 48(5): pp. 1873–1882.

49 Zhang, Y., R. Adams, and L.F. da Silva, Absorption and glass transition temperature of adhesives exposed to water and toluene. International Journal of Adhesion and Adhesives, 2014. 50: pp. 85–92.

50 Tucker, W.C. and R. Brown, Blister formation on graphite/polymer composites galvanically coupled with steel in seawater. Journal of Composite Materials, 1989. 23(4): pp. 389–395.

51 Miriyala, S.K., Galvanic Blistering in Carbon Fiber Polymer Composites. 1994, University of Rhode Island.

52 Sisneros, P., P. Yang, and R. El-Hajjar, Fatigue and impact behaviour of carbon fibre composite bicycle forks. Fatigue and Fracture of Engineering Materials and Structures, 2012. 35(7): pp. 672–682.

53 Greenhalgh, E., Mechanical Evaluation of Carbon-Fibre Reinforced Thermoplastic I-Beams. 1993. Defence Research Agency, Technical Report, 92071.

54 Kugler, D. and T.J. Moon, Identification of the most significant processing parameters on the development of fiber waviness in thin laminates. Journal of Composite Materials, 2002. 36(12): pp. 1451–1479.

55 Lightfoot, J.S., M.R. Wisnom, and K. Potter, A new mechanism for the formation of ply wrinkles due to shear between plies. Composites Part A: Applied Science and Manufacturing, 2013. 49(0): pp. 139–147.

56 Dodwell, T., R. Butler, and G. Hunt, Out-of-plane ply wrinkling defects during consolidation over an external radius. Composites Science and Technology, 2014. 105: pp. 151–159.

57 Mabson, G. and P. Neall III,, Analysis and Testing of Composite Aircraft Frames for Interlaminar Tension Failure. 1988, Rotary Wing Test Technology, p. 1988.

58 Miracle, D.B., S.L. Donaldson, S.D. Henry, et al., ASM handbook. Vol. 21. 2001, ASM International Materials Park: OH, USA.

59 Kassapoglou, C., Design and Analysis of Composite Structures: With Applications to Aerospace Structures. 2013, John Wiley & Sons, Inc.

60 Wang, J., K.D. Potter, and J. Etches, Experimental investigation and characterisation techniques of compressive fatigue failure of composites with fibre waviness at ply drops. Composite Structures, 2013. 100: pp. 398–403.

61 Chen, Z.M., R. Krueger, and M. Rinker, Face Sheet/Core Disbond Growth in Honeycomb Sandwich Panels Subjected to Ground-Air-Ground Pressurization and In-Plane Loading. 2015.

62 Hsiao, H.-M., S. Lee, and R. Buyny, Core crush problem in the manufacturing of composite sandwich structures: mechanisms and solutions. AIAA Journal, 2006. 44(4): pp. 901–907.

63 Martin, C., J. Seferis, and M. Wilhelm, Frictional resistance of thermoset prepregs and its influence on honeycomb composite processing. Composites Part A: Applied Science and Manufacturing, 1996. 27(10): pp. 943–951.

64 Renn, D., T. Tulleau, J. Seferis, et al., Composite honeycomb core crush in relation to internal pressure measurement. Journal of Advanced Materials, 1995. 27(1): pp. 31–40.

65 Shipsha, A. and D. Zenkert, Compression-after-impact strength of sandwich panels with core crushing damage. Applied Composite Materials, 2005. 12(3–4): pp. 149–164.

66 MIL-A-83376A (USAF), A.B.M.F., Sandwich structures, acceptance criteria, Notice 1, 30 July 1987.

67 Canada, T.S.B.o., Aviation Investigation Report, Loss of Rudder in Flight Air Transat, Airbus A310–308 C-GPAT Miami, Florida. 2007, The Transportation Safety Board of Canada: Gatineau, Canada.

68 Radtke, T., A. Charon, and R. Vodicka, Hot/Wet Environmental Degradation of Honeycomb Sandwich Structure Representative of F/A-18: Flatwise Tension Strength. 1999, Defence Science and Technology Organisation Melbourne (Australia).

69 Fogarty, J.H., Honeycomb core and the myths of moisture ingression. Applied Composite Materials, 2010. 17(3): pp. 293–307.

70 Hodge, A. and G. Dambaugh, Analysis of thermally induced stresses on the core node bonds of a co-cured sandwich panel. Journal of Composite Materials, 2013. 47(4): pp. 467–474.

71 Ley, R.P., W. Lin, and U. Mbanefo, Facesheet Wrinkling in Sandwich Structures. 1999, NASA Langley Research Center: Hampton, VA, USA

72 Nettles, A.T., Some examples of the relations between processing and damage tolerance. NASA Technical Report M12-2131 2012.

73 Campbell, F.C., Caution! Honeycomb Core can be dangerous to your health. Corrosion Reviews, 2007. 25(1–2): pp. 13–26.

74 Okada, R. and M. Kortschot, The role of the resin fillet in the delamination of honeycomb sandwich structures. Composites Science and Technology, 2002. 62(14): pp. 1811–1819.

75 Whitehead, S., M. McDonald, and R. Bartholomeusz, Loading, Degradation and Repair of F-111 Bonded Honeycomb Sandwich Panels-Preliminary Study. 2000, DTIC Document.

76 Garrett, R., Effect of Defects on Aircraft Composite Structures. 1983, DTIC Document.

77 Yang, P., S.S. Shams, A. Slay, et al., Evaluation of temperature effects on low velocity impact damage in composite sandwich panels with polymeric foam cores. Composite Structures, 2015. 129: pp. 213–223.

78 Hull, D., A unified approach to progressive crushing of fibre-reinforced composite tubes. Composites Science and Technology, 1991. 40(4): pp. 377–421.

79 Ostré, B., C. Bouvet, F. Lachaud, et al., Edge impact damage scenario on stiffened composite structure. Journal of Composite Materials, 2015. 49(13): pp. 1599–1612.

80 Johnson, A.F. and M. David, Failure mechanisms in energy-absorbing composite structures. Philosophical Magazine, 2010. 90(31–32): pp. 4245–4261.

81 Petersen, D.R., R.F. El-Hajjar, and B.A. Kabor, On the tension strength of carbon/epoxy composites in the presence of deep scratches. Engineering Fracture Mechanics, 2012. (90): pp. 30–40.

82 Shams, S.S. and R.F. El-Hajjar, Effects of scratch damage on progressive failure of laminated carbon Fiber/epoxy composites. International Journal of Mechanical Sciences, 2013. (67): pp. 70–77.

83 Chakravarthi, D.K., V.N. Khabashesku, R. Vaidyanathan, et al., Carbon Fiber-bismaleimide composites filled with nickel-coated single-walled carbon nanotubes for lightning-strike protection. Advanced Functional Materials, 2011. 21(13): pp. 2527–2533.

84 Qiu, Q. and M. Kumosa, Corrosion of E-glass fibers in acidic environments. Composites Science and Technology, 1997. 57(5): pp. 497–507.

85 Mouritz, A.P. and Z. Mathys, Post-fire mechanical properties of glass-reinforced polyester composites. Composites Science and Technology, 2001. 61(4): pp. 475–490.

86 Marty, P.N., N. Desai, and J. Andersson. *NDT of kissing bond in aeronautical structures.* in *16th World Conference on NDT.* 2004.

87 Adams, R. and P. Cawley, A review of defect types and nondestructive testing techniques for composites and bonded joints. NDT international, 1988. 21(4): pp. 208–222.

88 FAA, Composite aircraft structure, in FAA, Policy Statment, PS-AIR-20-130-01. 2014. 2014, Federal Aviation Administration.

4

Inspection Methods

4.1 Introduction

As described in Chapter 3, defects can develop as a result of faults in incoming materials, created during manufacturing or repair, or develop in service. Defects may not necessarily degrade mechanical behavior. The process of dispositioning a defect requires characterization of the defect (e.g. size, location within laminate, number of occurrences, etc.), an intimate knowledge of the composite structure, and an understanding of whether or not the defect can grow when the structure is exposed to service loads and environments. If the type, size and quantity is outside the published allowable limits, the responsible engineering organization will decide whether the structure is (i) acceptable as-is in the manufactured conditions, (ii) acceptable after repair, or (iii) not acceptable and should not be put in service. Proper defect characterization is critical to determining the actual effect. Note in all cases, inspection methods discussed in this chapter are not substitutes for proper manufacturing procedures.

Composite structures often present unique inspection challenges due to the non-uniform methods of construction (e.g. different layup, fiber and polymer material systems, production methods, etc.). Pass/fail criteria must be uniquely created for each inspection method with each material system, manufacturing method, and design detail. Second, many features cannot be nondestructively inspected after the part has been cured using currently available commercial methods. These features include bond integrity, laminate orientation, ply count, or any features that are closed-out in a bonded assembly. Experimental methods continue to evolve, but many of these are limited in application and have been only demonstrated in the lab environment conditions. The choice of the inspection methods presented in this section depends on whether destructive or nondestructive methods are desired. There may be several techniques that can apply to the same defect providing different levels of detail as shown in Table 4.1. Table 4.1 presents a mapping of the inspection methods relative the

Composite Structures: Effects of Defects, First Edition.
Rani Elhajjar, Peter Grant and Cindy Ashforth.
© 2019 John Wiley & Sons Ltd. Published 2019 by John Wiley & Sons Ltd.

Table 4.1 Common inspection methods for composite applications (post-cure).

	Mechanical vibration	Visual and enhanced visual NDT	Electromagnetic radiation (X-ray, gamma and neutron)	Optical methods	Strain measurement	Thermography
Typical methods	Ultrasonic: Thru-transmission, Pulse-Echo, Back scattering, Spectroscopy, Acousto-Ultrasonics, Bond testers, Low Frequency Vibration	Visual, Film Verification, Leak Testing[a]	Radiography, (Film, Computed, Digital)[b], Computed Tomography, Compton Scattering	Shearography, Profile meters and Imaging[c], Digital Image Correlation; Near-Infrared Hyperspectral Imaging	Strain Gages, Fiber Bragg Grating, Digital Image Correlation	Active and Passive Thermography, Pulsed Thermography, Thermoelastic Stress Analysis
Laminate defects						
Aged material						
Blisters	◆◆◆		◆◆◆ [1]			
Contamination or inclusions	◆◆	◆	◆◆ [1]	◆◆◆ [2, 3]		◆
Cuts, scratches or crushing	◆◆◆	◆	◆◆◆	◆◆◆ [2]		◆
Galvanic corrosion	◆◆◆ [4]		◆◆◆ [4, 5]			◆◆ [4]
Delamination (from impact or manufacturing)	◆◆◆ [6–8]		◆◆◆	◆◆◆ [2, 9]	◆ [10]	◆◆◆ [11]
Fiber damage	◆ [7]		◆◆◆			
Fiber/matrix debonding	◆ [12, 13]				◆ [10]	◆◆
Fiber misalignment or wrinkles	◆ [14–16]		◆◆◆ [1, 17]	◆ [2, 18–22]		◆ [23]

Defect	Markers / References
Incorrect fiber orientation; missing plies	◆ [24] ◆◆◆ [1]
Mismatched parts	◆ [13] ◆◆ ◆◆
Micro-cracks or crazing	◆ ◆◆ [1] ◆◆ [2] ◆
Moisture absorption	◆◆ ◆◆
Ply drops and gaps	◆ [3] ◆ [3]
Residual stresses and dimensional conformance	◆
Resin migration; un-even fiber volume fraction; thickness variations	◆ ◆◆ [1] ◆◆
Matrix degradation due to high temperature exposure	
Matrix degradation due to curing errors	◆ [25–27] ◆ [28]
Matrix degradation due to resin mixture errors	◆ [25]
Voids; porosity	◆◆◆ ◆◆◆ [17, 29–31] ◆◆◆
Sandwich structure defects	
Over-expanded; blown core	◆◆◆ [1] ◆◆◆ [2, 9]
Core crushing; movement	[6] ◆◆◆ [1] ◆◆ [2, 9]
Core splice spacing	◆◆ ◆◆◆ [1] ◆◆ [2, 9]
Incorrect or variable core thickness	◆◆ [1]

(continued)

Table 4.1 (Continued)

	Mechanical vibration	Visual and enhanced visual NDT	Electromagnetic radiation (X-ray, gamma and neutron)	Optical methods	Strain measurement	Thermography
Water entrapment in core	◆◆		[32]			◆◆
Incorrect density or density variations	◆◆		◆◆ [1]			◆◆
Misaligned nodes; unbonded nodes; mismatched nodes	◆◆◆ [33]		◆◆◆ [1, 32, 33]			
Core corrosion	◆ [4]		◆◆◆ [4] ◆ [34]			◆◆ [4]
Facesheet wrinkling, pillowing, or orange peel			◆◆ [1]	◆◆ [2, 9]		◆◆
Facesheet dents	◆◆ [35]	◆	◆◆	◆◆ [2, 9]		◆
Facesheet/core disbond	◆◆ [35, 36]		◆◆ [1, 5]	◆◆ [2, 9]		
Defects in adhesive fillet			◆◆ [1, 5]			◆ [37]
Edge-closeout defects			◆ [1, 5]			
Fastened joints						
bearing; hole elongation or out of round holes; missing fastener; misdrilled holes		◆	◆◆◆ [1, 5]			
Fastener seating		◆◆ ◆	◆◆◆ [1]			
Fastener over-torque or under torque						
Porosity near fastener	◆ [38–41]		◆◆◆ [1]			

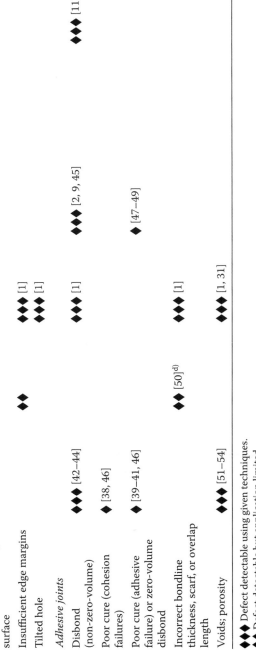

Resin-starved bearing surface

Insufficient edge margins

Tilted hole

Adhesive joints

Disbond (non-zero-volume)

Poor cure (cohesion failures)

Poor cure (adhesive failure) or zero-volume disbond

Incorrect bondline thickness, scarf, or overlap length

Voids; porosity

◆◆◆ Defect detectable using given techniques.
◆◆ Defect detectable but application limited.
◆ Technique under development in lab environment and showing promise.
Blank – Lack of reports or information in the literature about possible methods.

a) Microscopy is considered a destructive test. Many of the defects listed can be examined by microscopy if additional details are desired.
b) New methods can resolve the issue with defects that were typically not detected before, e.g. delaminations parallel to beam. Computed radiography is similar to film, but using digital replacement to film. Digital radiography enables real-time capture.
c) Methods are applied during layup stage to capture these defects before cure.
d) Verification or pressure film methods are performed prior to bonding or as follow-ups for investigation.

defects discussed in the previous chapter. Methods are classified relative to how common they are in practice. Experimental approaches that may be promising in limited application, or those that have been demonstrated in the lab environment, are also indicated separately.

In cases where features cannot be inspected, careful record keeping is essential and process control is of utmost importance. Record reviews can check features such as cure cycle temperature and pressure, and storage conditions.

The first part of this chapter describes the nondestructive inspection testing methods (NDT) that can be used to evaluate the properties of cured composite parts or assemblies without causing damage. The second part of this chapter describes characterization and destructive tests that can be employed to characterize uncured material or cured part quality. These tests can be used as quality control checks at receiving inspection or during the manufacturing process. These methods may also be called on when other inspections may indicate an issue. If used during production, these tests are typically performed on traveler panels/coupons or composite part prolongs (extended sections of parts that are manufactured for testing purposes).

The choice of NDT procedure is important to achieve a balance between the required inspection resolution and economic considerations. There are various limitations to each method. For example, some methods require water which maybe a problem if the structure is susceptible to moisture absorption, and other methods may limit the sample size. Furthermore, a preferred inspection method may not always be possible due lack of access.

Prior to the nondestructive inspection, calibration standards are created using the same layup and material as production parts. The calibration standards will contain defects to be captured such as disks of release film of different sizes to simulate delaminations and disbonds. Care should be exercised when manufacturing these defects to match the actual condition of the defect in practice. The required defect size for each NDT calibration standard can be established based on a guided assumption that a defect no larger than this critical size will be present within the structure. One approach to increase the confidence in an inspection method is to develop probability of detection (POD) curves [55, 56]. Using this approach, it is possible to determine for a given defect size, what the POD is going to be under defined variables such as lighting, inspector training, equipment, etc. Knowing that defects can be missed in inspection is closely tied to the design process. The composite design needs to account for the largest defects that have a high probability of being missed in the inspections.

NOTE: POD curves are not to be used to define the smallest defect that can be found. They should be used to find the largest defect that is missed.

4.2   Mechanical Vibration NDT Methods

4.2.1   Low Frequency Methods

4.2.1.1   Tap Testing

Tap testing refers to the technique of lightly tapping the composite structure or bonded joint with an instrumented hammer, coin or similar source to produce an acoustic wave within the material. Comparing the acoustic response from the composite to a reference area that is not defective allows the inspector to make judgments regarding the integrity of the structure. Typically, the well-consolidated or fully bonded areas will provide an even pitch sound, compared to an area with delaminations or disbonds that will produce a dull or muffled sound. An instrumented tap hammer has the benefit of some quantitative indication of the local stiffness (as measured by the impact pulse width with the use of computerization) [43] and reducing the subjectivity of the coin tapping approach (dependent on ambient sound levels and human hearing).

The tap test is limited by thickness. As the thickness increases, the difference between the response of "good" and damaged structure will get smaller. The tap test method is therefore not suitable for deep defects and should be limited to investigation of areas near to the surface. Developments in this field include Computer Aided Tap Testing (CATT), which facilities the generation of C-scans in field environments [57, 58]. The system consists of three components: the magnetic cart for mechanized tapping, the electronic circuitry for acquiring and processing the accelerometer signal, and software for displaying results [59]. The local stiffness can be measured by the contact duration of a vibrating hammer, which can be plotted as a function of position on the surface. Damage in the structure lowers the local stiffness and lengthens the contact time. The tap test can be followed by another NDT method (typically pulse echo ultrasonics) to more accurately determine the size and location of the defect. The benefit of this method is the low frequency of the input signals, localized approach, and speed of accomplishment. It is also attractive in situations where the composite is moisture sensitive and water-assisted ultrasonics is not an attractive approach because no coupling fluid is required between the transducer and the structure [42].

4.2.1.2   Mechanical Impedance

The physical basis of the impedance method is like the coin-tap test but using a continuous sinusoidal signal from a piezoelectric transducer rather than an impulse. The output and received signals are compared to determine the phase and amplitude correlating to changes in the structural impedance. It has been shown that defects such as disbonds and delaminations may be visualized as a spring beneath which is the undamaged structure, the spring stiffness being infinite if no defect is present [44]. The technique is most sensitive when the defect is close to the surface and the base structure is relatively stiff. In stiff structures (higher sensitivity), a skin-core disbond 10 mm in diameter can be found under a honeycomb structure with 0.5 mm thick carbon fiber reinforced polymer (CFRP) skins, but that with 1 mm thick skins a defect 20 mm in diameter is close to the margin of detectability [36].

4.2.1.3 Membrane Resonance

The membrane resonance method involves applying excitations in a wide band in the frequency range from 0.1 to 10 kHz using a piezoelectric transducer for both the transmitter and receiver at the same point. The method can be used to detect local delaminations and disbonds and does not require the use of couplants like those required in ultrasonics. Traditional Ultrasonics as the membrane, air coupled, and dripless boublers are all part of the ultrasound methodology. The method is most sensitive if the membrane resonance of the layer above the defect is in the frequency range of the excitation used. At the resonant frequency of the sublaminate, the vibration amplitude will be greater than in a region that is not affected by the defect. It is limited by depth but the performance at defect depths above about 4 mm is superior to that of the commonly used mechanical impedance and coin-tap tests [8]. It can also be used to detect bond defects between composite and metallic assemblies and to check for thicknesses.

The disadvantage of low-frequency methods is restricted to flaws near the surface. In thin laminates the results are less sensitive due to the overall lower stiffness. The contact pressure can also affect the measured results from all methods since the impact velocity may not always be controlled. In the resonance method, the distance of the coil from the surface will affect the measurements. Some commercial systems use springs to apply a constant pressure to the probe during measurement.

4.2.2 High Frequency Ultrasonic Methods

4.2.2.1 Pulse Echo and Thru-Transmission

In ultrasonic methods, a high frequency transducer (500 kHz to 20 MHz) introduces an ultrasonic wave into the composite part. The same or second transducer is used to interpret a transmitted or reflected signal. Pulse-echo or through-transmission methods are typically used. Coupling agents are typically required and may use water squirters, a surface gel, or immersion methods to improve transmission of the ultrasonic waves. The method is highly sensitive to delaminations, voids, and porosity and is able to locate them deep within the laminate. An A-scan is a measurement showing the time, distance, and amplitude of the received ultrasonic transmission signals. Using this method, it is possible to identify porosity and delaminations in composite laminates. C-scans are a two-dimensional representation of multiple A-scans, providing an image of the entire part.

At high levels of defects, accurate characterization becomes increasingly difficult due to large attenuation and reflections. For a given frequency, the void content is directly related to the attenuation in typical fiber reinforced composite materials. Stone and Clark [51] expressed the total measured attenuation A_T, as:

$$A_T = A_f + A_t + A_b$$

where A_f and A_b are associated with the loses due to the reflection from the front and back surfaces of the specimens (and not influenced by the condition inside the panel), and A_t represents the losses from the specimen itself. For good specimens with no porosity, the attenuation from the specimen was found to be around 3 dB. It is possible to compute theoretical values for A_f and A_b from the following:

$$A_f = 20\log_{10}(Z_w + Z_c)/2Z_c$$

where Z_w and Z_c are the impedances of the water and the composite respectively. For void contents greater than 10%, the total attenuation displays a parabolic relationship with void content when using higher frequency transducers (e.g. 10 MHz) and a linear relationship cannot be assumed. A strong relationship between the interlaminar shear strength and porosity has also been documented and observed (see Chapter 5).

4.2.2.2 Pulse Echo Ultrasonics

In the pulse-echo method, the same transducer is used to send and receive the ultrasonic signals. If there is no defect observed, the only reflection will be from the back-wall of the specimen (opposite side of thickness). Any defects will cause reflections to appear on the measuring scope between the initial pulse and back-wall reflection. Using this method one can observe the front wall, back wall, and reflections from defects in the part. Using time of flight measurements, the location of the defect (or reflector) can be located within the thickness.

4.2.2.3 Through-Transmission Ultrasonics

In the through-transmission method, two transducers are used with one sending the UT and the other receiving the sent signal. The transmitted signal from the specimens is adjusted so that the only losses are from the specimen and any attenuation in the response is due to the transmission losses in the specimen. In the C-scan, the map is used to associate a color (can be used for different defects but especially relevant to porosity). For example, it is possible to identify areas of constant porosity levels. The surface losses are typically not affected by the condition of the composite laminate and are independent of the part thickness [51]. The transmission loss, A_t may be assumed to increase linearly with the plate thickness, t and the absorption coefficient α (db mm^{-1}) dependent on the internal condition of the laminate ($A_t = \alpha t$).

4.2.2.4 Phased Array Ultrasonics

Phased array (PA) ultrasonics takes advantage of an array of ultrasonic sensors to interrogate through the thickness or large areas of the structure [60]. In contrast to the traditional method of the single point, a phased array probe system can be focused and swept electronically without moving the probe. The PA probe consists of many small elements, each of which can be controlled and pulsed separately. The elements can be pulsed at various times and at different frequencies. The multiple waves can be timed to add up to one single wave front traveling at a set angle.

4.2.2.5 Ultrasonic Spectroscopy for Zero-Volume Disbonds

Although not yet used in production environments, this method shows promise for evaluating zero-volume disbonds (also known as kissing disbond defects). The technique relies on the fact that solids vibrate at their natural frequencies when mechanically excited. In this method, elastic waves of constant amplitude and varying frequency are used to detect the sample's resonance. The peaks of frequency correspond to the natural frequencies. The natural frequency of a material depends on its elastic parameters. Thus, if a structure has a kissing disbond, its effective modulus will be different and it is then possible to interrogate the bondline condition by examining the condition to a reference part.

4.2.2.6 Ultrasonic Methods for Fiber Distortion

Ultrasonic measurements in composite materials for wrinkle measurements are complicated by the fact that fiber waviness defects do not typically create an impedance change to attenuate sound waves or cause reflections. Active research in the field is underway due to the popularity of this method in the industry to accurately capture the porosity and delamination defects discussed in the previous section. The research to date has not shown full resolution of the waviness morphology and is dependent on the material types and thicknesses inspected. For example, waviness may not always be uniform as it maybe localized to a few layers of the laminate or extend over large areas as described above. In addition, the waviness may also be coupled with porosity that complicates the reflection or transmission characteristics of the ultrasonic wave. Chakrapani et al. [15] proposed using ultrasound to measure the through-the-thickness waviness, and showed the periodicity of the waviness by correlation to C-scan measurements. Theoretical analysis of the longitudinal wave propagating through the material with the reflection coefficients was calculated at various locations. Wooh and Daniel [16] proposed using a discrete ray-tracing model based on elastodynamic ray theory. The varying anisotropy of the material causes the ultrasonic waves to travel along curved paths. The theoretical model was found to match the period of the fiber waviness. Scattering due to the fibers and porosity in the matrix makes this approach also challenging to implement. When the approach of using water as a couplant is not available, it is possible to use alternative methods such as air-coupled ultrasonics. When using air as a couplant, significant sound energy losses occur, and alternative methods are needed to minimize the energy losses at every stage of the measurement. In most cases, a dual transducer system is needed. Chakrapani et al. [15] used air coupled ultrasonic transducers with a two transducer setup to determine the aspect ratios of wrinkles at different depths in composite specimens used typically in wind turbines. The Rayleigh (surface) waves were generated using a 200 kHz frequency and were correlated to the aspect ratio of the waviness. The waviness with lower aspect ratios (producing larger degradation) are more clearly shown in the C-scans obtained using this approach (Figure 4.1).

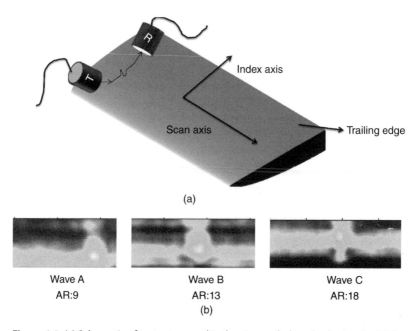

Figure 4.1 (a) Schematic of test setup used in the air-coupled method using Rayleigh waves to detect fiber waviness in a wind turbine blade (b) C-scan correlation to waviness aspect ratio, Chakrapani 2011 [15]. Source: Reprinted with permission from AIP Publishing LLC.

4.2.2.7 Guided Lamb Waves

In guided wave methods (typically less than 250 kHz), mechanical stress waves are propagated along an elongated structure while guided by its boundaries allowing waves to travel over large distances. The method can be used to inspect pipelines over large distances and can also be used in inspecting long prismatic or plate type structures. In composites the Lamb wave-based damage detection approach features the ability to inspect large structures with various defects while retaining coating and insulation [61–67]. Kessler et al. [6] successfully used Lamb wave methods for damage detection in composite materials using narrow graphite/epoxy specimens with pre-existing damage in the form of delamination, matrix-cracks, and through – thickness holes. Lamb waves have been investigated for barely visible impact damage (BVID) in sandwich structures of honeycomb core. It was found that residual deformations on the skin were almost linearly correlated with the depth of indentation and the crushed area within the honeycomb core [35]. In bonded joints guided lamb waves have been proposed to inspect for debonding [46, 68, 69]. Lamb wave interaction with in-plane fiber waviness in thin unidirectional composite laminates has been investigated by the aspect ratio of the defect and shear angles [14]. Air coupled ultrasonic transducers were used to generate the fundamental anti-symmetric mode (A0), which showed a reduction in velocity when interacting with in-plane waviness. The disadvantage of Lamb wave methods is that they require an active driving mechanism to propagate the waves, and the resulting data can be more complicated to interpret than for many other techniques [6]. Compared to other global frequency methods such as modal analysis, the Lamb wave methods can be responsive to local defects. Detection of multiple damage sites is issue of current research investigation. Complex scattering phenomena in the composite complicate the interpretation of Lamb wave signals, and distinguishing such signals remains problematic [70].

4.2.2.8 Air-Coupled Ultrasonics

Air-coupled guided wave measurements are based on the mode conversion from a plane ultrasonic wave in air to guided waves within the composite specimen, which takes place for angles of incidence satisfying the relation [71]:

$$\sin \theta = \frac{V_\alpha}{V_g}$$

where θ is the angle of incidence, V_α the velocity of sound in air and V_g the phase velocity of the guided mode. The radiation of the ultrasonic field in air by the propagating guided wave enables the use of a single-sided configuration. Application of this method requires adjustment of the incident angle and distance from the specimen to achieve the maximum output signal for the guided wave. Studies have shown that the guided wave area scans of specimens without in-plane fiber waviness (in-plane) showed uniform stripe patterns caused by wave guiding due to the 0° rovings whereas in the areas with local in-plane fiber waviness, the straight stripes become wavy and interrupted thus allowing for a non-contact and single-sided detection of fiber waviness [72]. They also performed fatigue-testing showing that stiffness degradation in specimens due to fatigue damage corresponds to a decline in the guided wave velocity in the specimens.

4.2.3 Acoustic Emission

Acoustic emission refers to the technique involving the analysis of the released energy in the form of stress waves propagating to the surface of the structure. The redistribution of stress in the structure due to an internal or external source like loading, damage, or temperature changes results from these stress waves. These signals can be detected by AE transducers containing a piezoelectric crystal. Some of the common measurements recorded include information on the time of arrival, frequencies, and amplitudes of the emitted pressure waves. The acoustic emission signals hold a large amount of information in the waveforms captured. The count information is processed by adding the number of times the voltage exceeds a set threshold. Energy content is usually processed by measuring the area under the amplitude-time curve (or envelope) for each emission [73].

Acoustic emission has also been described in numerous studies for application to composite pressure vessels [7, 74–78]. American Society for Testing and Materials (ASTM) E1067 describes a procedure that can be used on composite pressure vessels using the Felicity and Kaiser effects. Intermittent load hold-stress schedule is a quick test that can be used to identify severe accumulated composite damage using the Felicity ratio. The Kaiser effect establishes that in normal conditions acoustic emissions do not occur at loads not exceeding the previous maximum load level when the material undergoes repetitive loading patterns (Figure 4.2). The felicity effect is the defined as the presence of detectable acoustic emission at a fixed predetermined at a sensitivity level at stress levels below those previously applied. Whereas the Felicity ratio is the ratio of the stress at which the Felicity effect occurs to the previously applied maximum stress [78]. Applying the periodic unloading cycles it is possible to study the felicity ratio and the load levels at which the Kaiser effect breaks down. Investigators typically recommend focusing on the following AE parameters when determining the structural integrity of a pressure vessel: load at onset of emission, total emission, emission during load hold, high amplitude events, and the Felicity effect [78].

Reliability of the results of the acoustic emission depends on the compatibility of the hardware used; mainly the signal itself, the sensor and the pre-amplifier [79]. Amplifiers can have a high pass, low pass or a band pass type filters and are used to amplify and filter the AE signals coming directly from the sensors

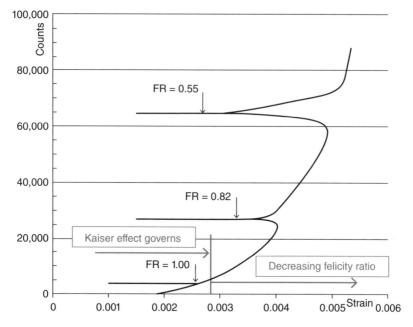

Figure 4.2 Acoustic emission on fiber glass composite showing Kaiser and Felicity effect [78]. Source: Creative Commons CC BY-ND 3.0.

or from a preamplifier before they could be processed. Other important aspect in using AE is source location of the events attributing to AE wave evolution. Conventional methods rely on the presence of more than one sensor and are based on measuring signal arrival time either by the zonal method, which uses the first sensor that detects the signals to specify the primary zone, or by the triangulation method, which measures difference in arrival time of the signal to an array of sensors [80]. Recent techniques are waveform based and include the cross correlation technique adopted by Ziola and Gorman [81].

Acoustic emission has been used to evaluate the manufacturing defects in composite structures. Dzenis and Saunders [82] applied acoustic emission to differentiate between pure and mixed mode fatigue fracture in adhesive composite joints by using transient AE signal classification by a computational panel recognition analysis. They found that signals from mixed mode failures are closer to failures from Mode-II signals in the lap joint test and that AE signals for pure Mode-I and II tests are not easily distinguished. Godin et al. [83] used AE to predict the fatigue life of ceramic composites during static fatigue loading. Similarly, Biermann et al. [84] measured and evaluated the AE from metal matrix composites (MMC) during cyclic deformation to monitor the initiation and propagation of cracks and Schiavon et al. [85] used AE and mechanical parameters to study the fatigue damage of one-dimensional glass/epoxy composites. Awerbush and Ghaffari [86] monitored the AE and various AE intensities including amplitude, duration, counts, and energy on the tension–tension cyclic fatigue loading of notched unidirectional graphite/epoxy AS/3501-6 specimens to study the emissions associated with matrix splitting in the specimens. Roy and Elghorba [87] on the other hand, monitored the interlaminar failure in Mode-II under monotonic and fatigue loading conditions by acoustic emission and used AE to investigate glass fiber/epoxy composite for damage development during monotonic and cyclic Mode-II loadings of DCB specimens.

Qamhia et al. [13] studied the acoustic emission responses from commonly observed manufacturing defects seen in composite structures. The continuous carbon fiber/epoxy composite laminates were manufactured containing varying levels of porosity and fiber waviness. AE intensities for counts, cumulative energy, and timing of events accompanied with mechanical parameters were used to differentiate the effect of manufacturing defects on the behavior of composites during unidirectional axial loading and the severity of each defect on composite properties. The results show that the porosity defects produced higher counts and energy releases than the waviness specimens but with less impact on the maximum stress. Coupling of the two defects increases the difficulty in identification of the source, thus it was not possible from the results to determine the extent of porosity or wrinkling when both were present at the same time.

4.3   Visual and Enhanced Visual Methods

4.3.1   Visual Inspections

Visual inspections are typically the first approach used to inspect for pre-cure inspection. Visual inspections are also used for surface flaws on cured laminates such as scratches, gouges, and nicks, and resin-starved or resin-rich areas. Optical aids, such as mirrors, boroscopes, digital cameras, or microscopes, can be used to improve the POD. In some translucent composite materials, visual inspections can find flaws a few mm below the surface. It is a subjective method in most cases.

In-plane fiber waviness on the surface plies can usually be observed using visual inspection with the proper lighting conditions. For out-of-plane fiber waviness, a cross-sectional analysis of a polished cross-section can yield important information about the defect levels. Sometimes shining a white or fluorescent light at an angle can reveal changes in the surface contrast indicating resin pockets that can correlate to the presence of out-of-plane fiber distortions.

4.3.2   Verification or Pressure Film

When co-bonding or secondary bonding applications are considered, visual inspection can also be performed on a sacrificial adhesive film using the verification film method (e.g. Verifilm™ by Cytec). The use of Verifilm essentially involves part layup, consolidation, and cure, but without adhesive bonding. In other words, a test part is laid up with a very thin (typically 0.001-in. thick) fluorinated ethylene propylene (FEP) film placed above and below the adhesive film layers, to prevent the adhesive from sticking to the prepared or "faying" metal/composite surfaces. The layup is processed normally, but then disassembled after cure (because the film prevented adhesion). The adhesive layer is carefully examined for indentations that indicate bondline defects such as improper fit and areas of resin migration. Bondline thickness is also measured.

Similarly a pressure distribution film (e.g. Fujifilm Prescale) can be used to indicate the pressure in the setup process before initiating the cure. Once preliminary setup has been formed, the pressure film can be used to troubleshoot production issues to detect areas of pressure non-uniformities. Also during production, pressure film can help guard against process deviations and can help guide decisions on where tooling or vacuum bagging setup adjustments are necessary.

4.3.3 Leak Testing

In the leak testing method, the visual inspection is aided by using a hot water tank to visualize air escaping. In this process, the composite part (usually a sandwich structure) is immersed in a hot water tank. As the part is immersed in the hot water tank, the air in the structure will expand and will escape as bubbles from any disbonds or edge close out cracks in the structure. The bubbles are visually observed and follow-on inspection with other methods maybe performed to further quantify the defects.

4.4 Electromagnetic Radiation (X-Ray, Gamma, and Neutron)

4.4.1 Radiography

Radiographic inspection (RI) is a noncontact inspection method which uses short wavelength electromagnetic radiation (X-rays or gamma rays) or beams of atomic neutron particles. Compared to the wavelength of visible light (6000 Å), X-rays have and gamma rays have shorter wavelengths (1.0 Å and 0.0001 Å, respectively) enabling them to penetrate materials that light cannot [1]. X-ray radiography can be used to probe local packing and orientation distributions in composites. Gamma and neutron waves can penetrate deeper into the materials. Radiographic testing involves exposing a media to X-ray radiation that has penetrated the composite specimen or structure of interest and subsequently developing an image from the media. An X-ray image intensifier and a high-performance television camera are required during the inspection [88]. The method can be used to detect obscure defects, such as lack of tie-ins, voids, and short cores. Depending on the density of the constituents, different regions will have different absorbance. The transmission of the radiation is improved with low energy and larger feature sizes but high energy is required to penetrate thick structures. Detection of small variations within the specimen with spatial details of 0.50 to 1.25 mm (0.020 to 0.050 in.) have been reported [89]. Some of the composite meso-structures (length scale typically from a few microns to centimeter range) can be determined by coupled radiography and de-ply methods [90]. In one of the hybrid methods, the laminate is penetrated with a solution of zinc iodide and subsequently the resin is pyrolysed in a furnace. This would be a form of destructive test since the penetrant cannot be removed. After pyrolysis, the individual laminas may be separated, and radiographs of these lamina provide a three-dimensional map of the damage pattern. Zinc iodide has a high radiation opacity and is an excellent penetrant that can be added to improve the contrast. The penetrant should have a path to enter the composite to allow the inspection during the X-ray scanning phase. By assessing the radiographs, it is possible to make an assessment of local fiber orientations and other forms of internal damage. Generally, the polymer matrix materials and carbon fibers have a low-absorbance and so low energy X-rays are used. It is difficult to identify the two phases to calculate the fiber fraction, but it may be possible to use in other non-polymeric systems dependence on the ratio of the absorbance of the constituent materials. The X-ray radiation absorption in the specimen, results in the appearances of various shades of gray. It is also possible to focus the radiation beam to obtain very good resolution in the few micron range. Delaminations or other defects that do not have depth in the beam direction cannot be typically detected.

4.4.2 Computed Tomography

High-resolution X-ray computed tomography (CT) offers a unique approach for characterizing porosity in fiber-reinforced composites. This technique uses highly charged X-ray particles to penetrate the sample of interest from different angles of scanning the part to produce 3D microstructure reconstructions. It is important to note, that the current technology does not render this method for field application since core specimens have to be extracted from the original structure. CT scanners for smaller parts such as the MU2000 CT system can inspect smaller components. For example, in Figures 4.3 and 4.4, the CT system can produce a 3D rendering of the thickness values in the carbon fiber bicycle frame showing a correlation between the minimum thickness and the location of the crack. Figure 4.5 shows application of CT for layer identification of a helicopter blade part identifying the layer geometry and distribution and sizes of voids.

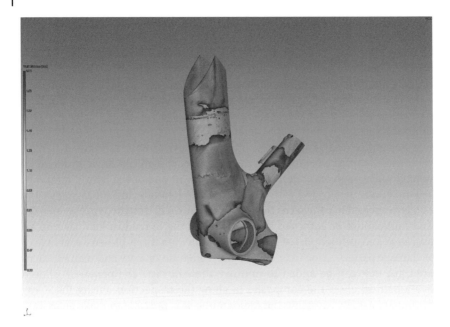

Figure 4.3 Thickness variation in a bicycle frame obtained using the MU2000 CT system. Note the crack near the region of minimum thickness. Source: Reprinted with permission from YXLON.

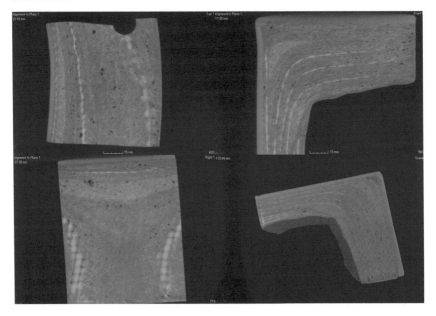

Figure 4.4 CT Scan of helicopter blade section that attaches to the hub showing porosity and laminate plies. The scans were produced using the Y.CT Modular with the 16″ digital detector array. Source: Reprinted with permission from YXLON.

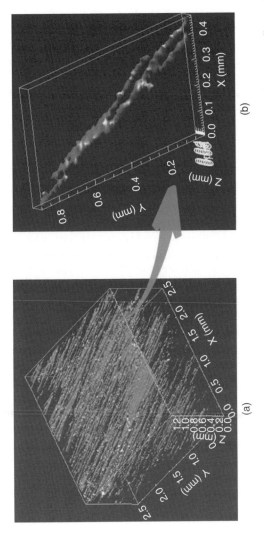

(a)

(b)

Figure 4.5 Void shape and distribution of porosity in carbon-fiber composite using X-ray computed tomography (a) overall porosity distribution (b) morphology of connected porosity pore, Yang 2014 [91]. Source: Reprinted with permission from Taylor and Francis.

X-ray synchrotron tomography, which uses radiation within the electromagnetic spectrum, has shown promising results in identifying porosity as well as damage growth studies [92, 93]. Previous studies have compared X-ray CT with several conventional methods for void content analysis in fiber reinforced polymer composites (FRP). It was determined that X-ray CT can yield more accurate results in determining void content when compared to ultrasonic testing and acid digestion [31]. CT is able to detect void and porosity within 0.5% whereas ultrasonic and acid digestion as accurate to within $\pm 1\%$. Damages in FRP such as micro-cracks, delamination, and fiber breakage can also be observed three-dimensionally in high resolution images [94, 95]. A method for accurate measurement of voids has also reported based on X-ray Computed Micro-Tomography, based on a sub-pixel contour generation for the average of the air and material gray values obtained in CT scans and correlated to the material density [17]. In addition to the volume percentage in the composite, further assessment may be needed to understand the morphology of the voids and porosity. The size, shape, and orientation of pores in carbon-fiber/epoxy composites has previously been investigated [[30], (Yang, 2012 #27)]. Some of the results indicate that pores align to the fiber direction and have a distinct aspect ratio of long extruded quasi-elliptical shapes (Figure 4.5). Stone and Clark [51] have also shown that for low void contents (less than 2%), the voids tend to be spherical in shape whereas for voids greater than this they tend to be of the interlaminar type and are caused by air trapped between the ply layers which is more flattened and elongated. McMillan [96] demonstrated the importance of understanding the morphology of the porosity defects. It was found that the void shape and percentage porosity levels play a significant role in the strength reduction; another critical factor is the variation of distance between the voids.

4.4.3 Compton Scattering

Compton scattering is the inelastic scattering of a photon by a charged particle, usually an electron. The back-scattered X-rays are elastically scattered by interaction with atoms in the composite, and are picked up in this one-sided inspection method. Specific advantages of X-ray scatter imaging include a flexible choice of measurement geometry, direct 3D-imaging capability (tomography), and improved information for material characterization [97]. The Compton scattering effect becomes dominant at higher energies of the beam (tube voltage larger than 100 kV).

The use of ionizing radiation creates a slew of issues with protecting personnel from harmful effects. The use of the penetrant is limited by the ability to have an entry path of the penetrant into the laminate. Scanning from multiple angles is necessary to achieve information about depth. The ability to use focused radiation allows the production of high quality images of the internal

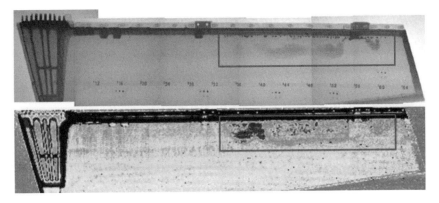

Figure 4.6 Comparison of neutron tomogram (top) and UT C-scan (bottom) from honeycomb composite rudder with moisture entrapment in core, Hungler 2011. Source: Reprinted with permission from Elsevier.

structure. It can also allow for inspection of thick composites where issues exist when using the other methods. On the right materials or when using enhanced methods, it is possible to determine information about fiber volume fractions and allows for detection of anomalies in the fiber features such as waviness.

4.4.4 Neutron Tomography

Neutron tomography as a type of CT involving the reconstruction of 3D images from neutrons detected after passing through the material from the source. Multiple planar images are used to reconstruct the object. Neutron imaging has been applied to investigate water ingression into the honeycomb cores. When water has made it into the core the neutron tomography method has been shown to differentiate between areas which the skin to core fillet bond has deteriorated compared to UT which can identify the areas impacted with moisture (Figure 4.6).

4.5 Optical Methods

4.5.1 Shearography

Shearography is a full field method that allows for determining the strain distribution over a large area to be observed. The output of shearography is a fringe pattern depicting the distribution of strain. It reveals flaws in materials by looking for flaw-induced strain anomalies which are translated into anomalies in the fringe pattern [98]. The method does not require special vibration isolation and can be employed in field/factory environments (Most frequently

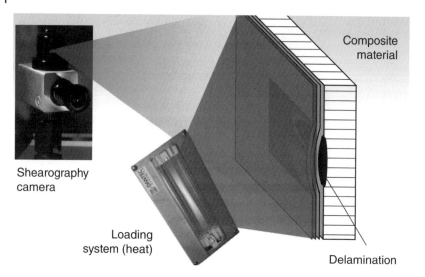

Composite
material

Shearography
camera

Loading
system (heat)

Delamination

Figure 4.7 Application of shearography on composite material. Source: Courtesy of Dantec Dynamics.

used in manufacturing at component level) [99]. In this method a coherent light and optical interference are used to provide information about the composite material by detecting localized stiffness changes. The slightest surface excitation leads to surface deformations from the internal flaws as shown in Figure 4.7. Applying very small changes in stress (by increasing the temperature or pressure) can alter the out-of-plane deformation compared with the rest of the structure. A laser light is used to detect sub-micrometer changes in the surface deformation. A reference image is obtained by using the scattered light projected onto an image plane by a special shearing lens with this lens shearing the image of the composite specimen in the plane of the lens. This allows interference to occur between the direct and sheared images. The method has been used in production and development in aerospace, wind turbine blades (Figure 4.8), and materials research [9]. Averaging and transient method variations have also been proposed for fine-tuned responses [100, 101]. Similar to other optical methods, the technique allows the inspection of large areas of the structure with a non-contact approach. The technique has been used in composite laminates of thicknesses of near 40 mm. Typically a mechanical or thermal approach is necessary to excite the structure and a second interference pattern is recorded. A fringe pattern is produced by using the reference and second recorded interference pattern of the images obtained with the structure in the stressed and unstressed states. A higher concentration of fringes implies increased strain concentrations as each fringe represents a strain value. The deformation near a defect will interact differently than the surrounding pristine

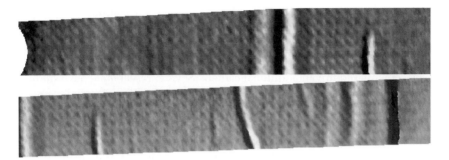

Figure 4.8 Wrinkles in turbine blade obtained from shearography test. Source: Courtesy of Dantec Dynamics.

or damage structure. This local deformation can be readily identified from the image obtained through the interferometer. The underlying defects result in a lower stiffness normal to the surface and the distortion produced by the local out-of-plane pressure reduction is hemispherical and this leads to the double bulls-eye interference pattern as shown in Figure 4.7. The method has been used to quickly identify wrinkles, BVID, delaminations, disbonds, porosity, and surface damage. It has also been reported to apply well to core damage and core splice joint separations. The technique can be qualitative for some defects due to the limitations on material thickness, proximity of defect to the surface, and provides no depth information directly like the pulse-echo ultrasonic method [102]. Research methods are being developed that use vibration and resonance to distinguish depth information [45, 103, 104]. Also in cases of application to woven materials, the yarn crimp effects can be significant because of the constraint effect of underlying layers and localized distortions from the weave pattern [105]. A major disadvantage of shearography (like thru-transmission ultrasonics) are the defects at the same planar location but at different depths within test samples will show up as only one defect [106]. Despite this, shearography can produce results very quickly compared to other techniques, e.g. ultrasonic Testing and X-ray methods [102, 106].

Most sensitivity in the shearography is obtained from out-of-plane changes in deformation but recent systems allow the separation of in-plane and out-of-plane components. Shearography systems can be built as portable units or mechanized for scanning large structures (see case study in Section 7.1). Rigid body movement of the composite structure or part can be accounted for in this method since shearography uses the deformation derivatives.

4.5.2 Digital Image Correlation (DIC)

The increased use of image correlation methods made possible by advances in theory and cost reduction of equipment have extended the use of such coupled methods for composite material characterization. In this method, surfaces are painted with speckles, then high-resolution cameras are used with image registration techniques to accurately track in-plane and out-of-plane displacements. The in-plane test setup typically involves one camera; whereas the out-of-plane displacements require an additional camera. Several researchers have studied the deformation and failure behavior of composite materials using digital image correlation (DIC) methods. Goidescu et al. [107] used a coupled DIC, infrared thermography, and X-ray tomography approach to study the failure of CFRP composites in a qualitative approach observing material anisotropy and failure aspects. Johanson et al. [108] used the Lagrangian strains from DIC to study the architecture and failure in discontinuous carbon fiber composites formed from randomly orientated bundles and to map the evolution of the failure. Kissing debonds have also been evaluated using DIC methods [47–49]. In filament-winding application, Henry et al. [20] used a two-dimensional DIC method to study the behavior in unidirectional and cross-ply laminates. The observations of strain on the free edge revealed that wrinkles caused elevated out-of-plane shear and through-thickness normal strains in regions eventually involved in the fiber micro-buckling failure process. Makeev et al. [109] tested thick section carbon-epoxy multidirectional laminates using DIC to measure the strains and used the nonlinear shear behavior to predict the failure load.

There is clear evidence of early failures in laminated composites during tension testing caused by wrinkles [110]. The results show interlaminar failures occur prior to the maximum load without a significant load drop that can be used to identify the early failure. Using optical video microscopy on the side of the specimen, it was possible to qualitatively identify the interlaminar damage. In a composite specimen this can clearly occur without a significant drop in the load-displacement response. This load drop in the response is normally used to study the onset of failure. However when there is no load drop observed in the load-displacement record, a non-conservative assumption of using the ultimate stress for characterization of the material system is typically used. Without the load drop it is not possible to accurately define the onset of failure in the composite material. Further, it is not always possible to monitor the specimens using a video microscope if the defect is not visible on the side (e.g. wrinkle does not extend to the edge) or if the specimen geometry precludes this from happening. Elhajjar [19] proposed a method to determine the limit point in composites loaded in tension using the out-of-plane displacement tracking capability using an image correlation method to determine the start of incipient interlaminar delamination in continuous fiber reinforced composite materials containing wrinkles (Figure 4.9). The method is based on using a two-camera 3D-DIC setup for observation of the surface of the composite material during a normal tension test for capturing of the in-plane and out-of-plane deformations. The maximum principal strain is used to identify the hot spot stress location for construction of an evaluation line. The z-displacement along this

z-Displacement (mm)

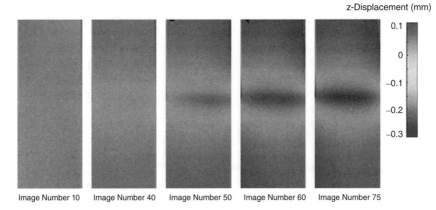

0.1

0

−0.1

−0.2

−0.3

Image Number 10   Image Number 40   Image Number 50   Image Number 60   Image Number 75

Figure 4.9 Progression of out-of-plane displacement in a CFRP/epoxy multidirectional layup $(0/45/90/-45/0/45/-45/0)_s$ containing a wrinkle defect loaded in tension. Out-of-plane displacements jumps occur at limit points, Elhajjar 2016 [19]. Source: Reprinted with permission from Elsevier.

evaluation line is then tracked as the specimen is loaded. Differentiation of the z-displacement field results in peaks that are subsequently used to identify the limit point of the composite material.

DIC is also able to provide strain estimates and shape information which may aid in the evaluation process. The method has been used to inspect blades as it is small and compact and can be easily mounted on wind turbine blades for example. The quantitative nature of the tests also allows for the ability to determine the criticality of the defects. In production, the strain distribution in critical areas of the composite structure (e.g. turbine blade) are compared to maximum strain values for pass/fail criteria. Real time correlation can be used to monitor a composite structure during fatigue testing to detect the extent of damage growth.

4.5.3 Hyperspectral Near Infrared Method for Resin Migration and Fiber Distortion

Chemometric data processing methods can be used to characterize the relationship between the spectra and the resin thickness. This information can provide the correlation to predict a point-by-point local resin thickness and thus produce thickness maps of the resin on a scanned surface of a composite structure. Recent advances in hyperspectral near infrared (NIR) sensing and data processing technologies have made real-time infrared methodologies a viable solution for accurate diagnostics and prognostics of composite structures. NIR imaging can accurately measure surface resin on composite materials from 125 to 2500 μm thick, detecting virtually all resin pockets, resin-filled

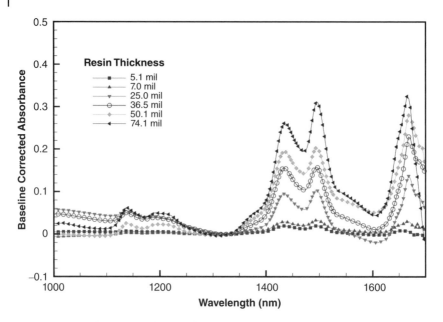

Figure 4.10 Baseline corrected near infrared spectra (1000–1700 nm range) for a range of epoxy resin thicknesses. A calibration is created to correlate the measured absorbance with resin thickness at each scanning point, Elhajjar 2016 [21]. Source: Reprinted with permission from Elsevier.

surface wrinkles, and other surface resin features [22]. The resin-rich areas on the surface are usually an indicator that wrinkles are impacting some or all of the plies in a laminate stack. Figure 4.10 shows the reflectance spectra obtained on resin standards with different thicknesses that were made on test coupons to mimic the composite structure with resin on its surface showing the different spectra achieved for each resin thickness [21]. In the push-broom NIR hyperspectral imaging approach, a line of pixels is measured and the sample is moved along a path that is perpendicular to the imaging line. The push-broom approach to imaging is capable of high spatial resolution to completely define the resin features by generating a 3D profile of the resin features on the surface of the composite structure. Commercial applications of this technique to aerospace structures has been reported in the literature [22].

4.5.4   Laser Profilometers and Image Processing

Laser Profilometers and cameras can be used to aid in positioning materials during the layup process. They are also used for visual inspection to investigate the presence of foreign objects, gaps, overlaps, and fiber uniformity during the manufacturing process by inspecting each ply as it is deposited on the tool

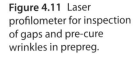

Figure 4.11 Laser profilometer for inspection of gaps and pre-cure wrinkles in prepreg.

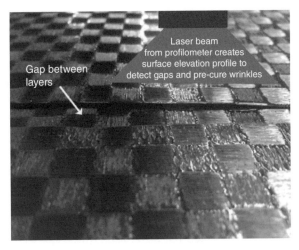

(Figure 4.11). The laser profilometer is a device that projects a laser onto a surface and measures the distance at each point, thus generating a 3D array of the surface. This array can be used to create a profile of the surface that can be used to detect foreign object debris (FOD) or irregularities with laying up the fibers such as gaps, overlaps, and fiber bridging. Modern systems for pre-cure inspections still under development promise to be mounted on the fiber depositing machine and can use feature recognition by using data from multiple sources such as cameras (for ply boundaries and tow-ends), laser profilometers (gaps, overlaps, bridging, and FOD), part programs, and operator input [3]. Challenges in these systems include the ability to contain the laser and have a more automatic defect recognition capability. A user interface can be used to identify defects and locate them on the actual part models and using laser projectors can also be projected back to the part surface.

4.6 Strain Measurement

Strain gages can be a very effective method for NDT of composites. The technique is very mature and there are a variety of foil strain gages available for a variety of situations such as determination of stress intensity factors and coefficients of thermal expansion [111–113]. The point limitation of this method is overcome by deploying many strain gages, the inspector can determine the likelihood of failure or if there was any overload or loss in stiffness. Fiber Bragg Grating (FBG) sensors have been used to monitor temperature and strain changes in composite materials during cure and service [114, 115]. The FBG has a longitudinal periodic variation of the refractive index such that the intra-core gratings are written using high intensity ultraviolet (UV) laser source. Wavelength of the light reflected at this fiber depends on the spacing between the refractive index variations. The light will be scattered in the regions of varying refractive indices. This can be coupled with neural networks and modal shape analysis to predict the delamination damage [116].

Applications have been seen in composite piles [117], bridges [118, 119], repair and strengthening applications [120], and in radome sandwich structures [121]. Some concern was expressed on the longevity of the sensors or in damage they can induce on the long term, but specimens tested after many years do not show damage due to the optic fibers [122].

Researchers at the National Wind Technology Center near Boulder, Colorado monitored the distributed strain using optical fibers while a wind turbine blade was manufactured and later cycled to failure [10]. Distributed strain measurements were taken on a carbon fiber CX-100 wind turbine blade with intentionally introduced out-of-plane waviness in the part. The team instrumented the blade using three surface-mounted fibers co-located with electrical foil gages. Static tests showed good agreement between the strain gages and surface-mounted fiber data, and the internal defects were clearly evidenced by the surface fibers. The blade failed at 1 968 000 cycles due to a defect on the high-pressure side of the blade. The fiber optic solution identified the critical defect by cycle 614 000, which correlates with the time that low-level acoustic emission was first detected from the blade. The sensors therefore predicted a crack or delamination was forming at the critical defect by 614 000 cycles, even though it was not visible on the surface of the blade until 1 950 000 cycles.

4.7 Thermography

In this method, flaws are monitored using an infrared camera after the introduction of heat into the system by either a heat source or through mechanical loading. This method is gaining popularity as infrared cameras are increasingly available in lower costs due to the introduction of microbolometer sensors. Thermography is through either passive or active methods.

In classical passive (or pulsed) thermography methods, a heat (or cooling) source is introduced as a pulse to the surface with the intent that it will interact with the defect. The defect will disrupt the flow of incoming heat as it conducts into the material. Surface defects are particularly sensitive to this temperature change. Pulsed thermography, used under pre-production conditions (such as during process start-up and optimization trials) has been effective in assessing the quality of structural adhesive bonds and the quality of the composite itself [123]. The performance is dependent on the heat source used. However, because the carbon-composites structure has different in-plane and out-of-plane conductivities, internal defects maybe obscured using this approach. Similar to the ultrasonic method, this approach can be used in pulse-echo form where the heat source and imaging camera are on the same side, or through-transmission where the camera and heat source are on opposite sides.

Active thermography methods are those where a cyclic stress is induced to the specimen – either due to induced vibrations such as using a fatigue machine, or by inducing resonance in the structure. Researchers have shown that areas in a composite structure containing local fiber waviness and loaded under compression tend to heat slower (leading to lower local temperatures) compared to defect-free structures [72]. This method requires the capability to apply repeated cyclic loads and to measure the development of the surface temperature over a certain cyclic range.

4.7.1 Active Thermography

4.7.1.1 Thermoelastic Stress Analysis

Thermoelastic Stress Analysis (TSA) is an active infrared method sometimes referred to as vibrothermography. In the TSA method, during cyclic loading and the presence of reversible adiabatic conditions, an infrared detector measures an un-calibrated TSA signal. Using a linear detector and assuming the Stefan–Boltzmann relationship relating the flux change to small changes in the surface temperature, in isotropic materials and for practical application, the thermoelastic stress analysis infrared (TSA-IR) signal (S) can be related to the first invariant of the stress, $\Delta\sigma$, through a calibration constant, k_σ ($\Delta\sigma = k_\sigma S$). Thus, in isotropic areas of the structure, one expects the TSA-IR response to be directly proportional to the first invariant of stress and is representative of the stress conditions at that location. The thermoelastic equation for an isotropic homogenous material loaded elastically and under adiabatic conditions yields $\Delta T = -\frac{\alpha T_o}{\rho C_p}\Delta(\sigma_1 + \sigma_2)$ where ΔT is the change in temperature, C_p is the specific heat at constant pressure, α is the coefficient of linear thermal expansion, and T_o is the reference temperature. This method has been successfully used to characterize the out-of-plane fiber waviness in a continuous carbon fiber/epoxy composite laminate composite system [23]. The TSA method was able to identify two different waviness profiles compared to the control. The TSA method shows promising results in highlighting the location of the induced waviness imperfections made in the laminates. The high infrared emissions from the epoxy rich areas are the enablers for identifying these zones (Figure 4.12).

4.7.2 Passive Thermography

In passive or transient thermography, the composite structure is exposed to a heat pulse in the form of a step function or sudden exposure. The heat can be

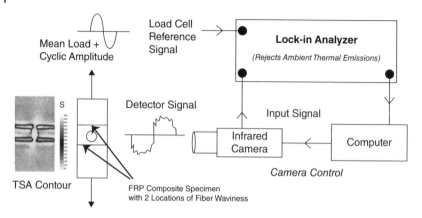

Figure 4.12 Schematic of test method for TSA method identifying resin pockets associated with out-of-plane fiber wrinkling.

applied using hot air guns, hot water spray, steam, heat lamps, or flash lamps near the surface. An infrared camera is set up to observe the structure as this signal is turned off – either on the same side as the source, or opposite to the thickness as in the thru-transmission method in ultrasonic inspection. This is becoming much more utilized in maintenance for inspection of flight controls and overcomes the dangers associated with X-ray. When a delamination is present, the temperature will rise more slowly compared to the intact laminate away from the delamination. Image analysis can be used to take time derivatives of the infrared images to obtain further analysis of the experimental results. The method is also particularly useful when applied in the context of heat blanket repairs to check for heat sinks and also to monitor the potential for overcure and undercure conditions when repairs are made. Pulsed thermography for nondestructive evaluation and damage growth monitoring of bonded repairs has correlated results with ultrasonic pulse-echo C-scan inspections and destructive testing show good disbond detection capability with an accuracy similar to that of ultrasonic inspection [124]. Pulsed thermography can be an effective technique for quantitative prediction of defect depth within a specimen [125, 126]. In some cases, the method showed that it can be used effectively in the detection of subsurface composite features (strut) located beneath 2–4 mm in a composite fiber composite [127]. The advantages of the technique are that it investigates rapidly large areas for surface or near surface defects and that it produces easily interpretable results [128]. Figure 4.13 shows a visual image of honeycomb panel with four impact areas outlined in pencil and a thermal image (right) where the impact areas can be seen [129]. Darker spots correspond to the damage produced. The impact energies for the four impacts were 1.35 J (1ft-lb) and 5.42 J (4ft-lb) for the top left and right and 2.71 J (2ft-lb) and 4.07 J (3ft-lb) for the bottom left and right areas respectively.

30 cm

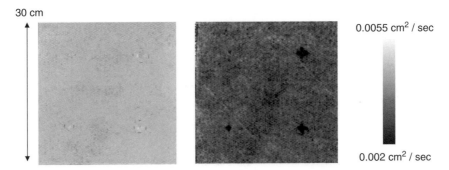

0.0055 cm^2 / sec

0.002 cm^2 / sec

Figure 4.13 Visual (left) and thermal diffusivity inspection image (right) of a honeycomb panel, Anastasi 2004 [129]. Source: Courtesy of NASA.

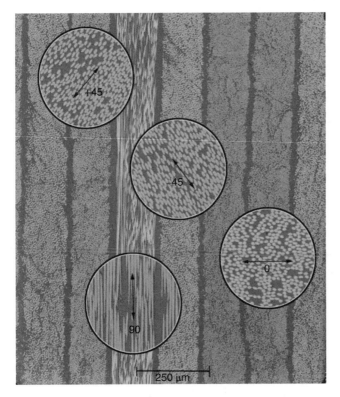

Figure 4.14 Bright field illumination microscopy from a carbon fiber composite showing ply angles Hayes 2010 [130]. Source: Reprinted with permission from ASM.

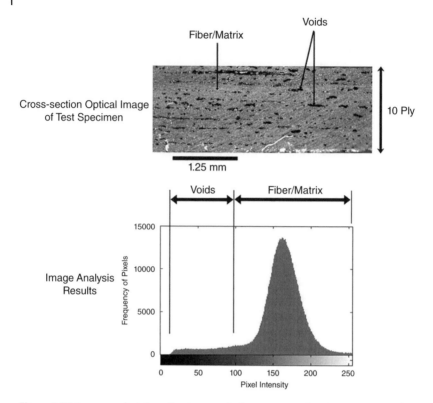

Figure 4.15 Image analysis to estimate porosity from cross-section images captured on digital scanners and analyzed using image analysis software.

4.8   Destructive Methods

The methods detailed in this section are typically used for incoming material, product development, or for quality control procedures. They can be used as secondary testing methods in case the NDT proves inconclusive or calls for further investigation.

4.8.1   Physical Tests

Table 4.2 shows some of the common physical and chemical tests that can be conducted on the constituents during raw material manufacturing or during receiving inspection of uncured or cured materials. These tests can also be performed on traveling specimens to confirm certain process steps are achieved. Traveler and test specimens should be processed using the same methods that

Table 4.2 Constituent and lamina level destructive tests.

Test property	ASTM test method
Resin content	D 3529, C 613, D 5300, D 3171
Resin volatile content	D 3530
Resin gel time	D 3532
Resin flow	D 3531
Fiber areal weight	D 3776
Infrared spectroscopy for polymer identification, contamination, and processing changes	E 1252, E 168
Differential scanning calorimetry or dynamic mechanical analysis for glass transition or processing effects on polymer chemistry.	E 1356, D4065, D4440, D5279, D 7028,
Fiber tests	D2343, D4018
Void content	D 3171, D 2734
Cured resin density	D 792

are used to manufacture the actual structure. While they serve a purpose they are no substitute for quality and process control. For example, a resin cured in an oven at one temperature will have a different property when processed in an autoclave at a different temperature. Thus, it is important that the material tests follow the actual process as much as possible. Incoming fibers and fabrics may also have material defects. Both glass and carbon fibers come in different grades and in varying tow-sizes. For example, glass can be available in E-glass or S-glass and carbon fibers can come in high-modulus or high-strength variety. They could additionally be broken or kinked during handling which may impact the properties of the final composite. The distribution of the fibers and bundling may also induce the formation of resin pockets or air gaps in the final composite part.

Overcuring or undercuring can occur whenever there is a thermosetting polymer material included in the process of manufacturing composites. When the optimal cross-link density in the thermosetting polymer does not occur in the composite, numerous properties of the resin can be affected. For example, the glass transition temperature, $\mathbf{T_g}$, modulus, hardness, conductivity, strength, and degradation temperatures are all affected. Other properties such as the coefficient of thermal expansion and moisture uptake will also be affected. Methods such test methods in Table 4.2 can be used to study the degree of cure and evaluate the changes that may have occurred.

4.8.2 Density and Porosity Measurements

Porosity is generally evaluated using physical methods such as resin digestion, optical microstructure assessment, or ultrasonic attenuation measurements. The methods can provide an average porosity level usually expressed in percentage form. With some methods it is possible to determine if the porosity is stratified or distributed, although in many cases, little is known whether this porosity manifests itself in a homogenous fashion or in some directional pattern. There is no single acceptable porosity value, but any porosity in the structure should reflect the porosity in the test specimens used to determine the mechanical properties. A value of 1% to 2% is considered typical in well-consolidated laminates. The void or porosity content, % can be determined from the following equation:

$$V = 100 - M_d \left(\frac{r}{d_r} + \frac{f}{d_f} \right)$$

where:

M_d = measured density,

r = resin, weight %,

f = fiber, weight %,

d_r = density of resin, and

d_f = density of glass

Note that the bulk density of the resin may differ from the actual density in the composite due to differences in the curing and molecular interactions between the resin and the surface of the reinforcement. The error should be recognized and is more significant if the true volume of voids is low. Furthermore, in semi-crystalline plastics (e.g. polyetheretherketone (PEEK) and PPS), the degree of crystalinity may significantly be affected by processing and may affect the density used in the calculation. Differential Scanning Calorimetry (DSC) or X-ray diffraction can be used to determine the degree of crystallinity of the polymer.

Matrix removal by acid digestion, ignition, or carbonization is often suggested as one of the destructive methods to determine the porosity content. Removing the resin and leaving the fibers or reinforcement unaffected allows the calculation of the fiber and matrix content as well as calculation of the percent void volume. In these methods, the density of the composite is measured and compared to a theoretical density of the constituents [131]. The density of the composite, resin, and the fibers are measured separately. The method assumes that the matrix removal process does not affect the fibers and minor changes in the reinforcement can be accommodated [132]. The ASTM test method also has a procedure to account for matrix and fiber content (but no void information) from a known areal weight and measured laminate thickness.

A relationship between the level of porosity and the tanδ peaks can be observed in Dynamic Mechanical Analysis (DMA) tests on CFRP specimens where a temperature ramp is used [133]. The test requires a small specimen that may be extracted from a specimen edge or from a traveler coupon. A significant shift in the T_g toward a lower temperature has also been observed with increasing porosity. The tanδ peak is usually associated with the glass transition temperature but can also be used for porosity content assessment [133]. It has also been observed that the *amplitude* of the peak of the tanδ increases almost linearly with the increase in the level of porosity, although the Tg itself is lowered, indicating that levels of porosity in the specimens might be characterized by this parameter. This result can be useful since a small specimen can be quantitatively used to indicate the level of porosity with environment and its detrimental effect on the mechanical behavior simultaneously.

4.8.3 Microscopy

Optical microscopy can be a very effective tool in failure analysis and quality control of composites when defects are considered. It can yield very important information about the microstructure and macrostructure of the fiber reinforced composite. Optical microscopy can be used for determining void content, ply counts, and fiber misalignments (both in-plane and out-of-plane). Microscopy can also be used to understand other factors such as fillers, matrix additives such as second phase thermoplastic toughening agents, fiber geometry, and packing. Two methods are commonly used in microscopy of composites: reflected light microscopy and transmitted light methods. A good image requires careful sample preparation and polishing the specimen to achieve the desired quality. Reflected bright field illumination (Figure 4.14) is particularly useful for imaging specimens for ply counts, fiber orientation verification, resin to fiber ratio determination, void studies, and micro crack investigation whereas sub-surface features such as microcracks can be obtained using dark field illumination [130]. Transmitted light microscopy is generally used only when the reflected methods yield limited results and more information is needed for failure investigation such as from lightning strikes, studying matrix fillers, localized analysis for contamination or areas where UV or other degradation needs to be examined.

Image analysis of the gray scale pixel values of a polished cross section can be used to distinguish the difference between fiber, matrix, and voids [54]. The method has been used successfully in the analysis of carbon fiber/epoxy composites [30]. Typically polishing is required using both 320 and 600 grit silicon-carbide abrasive disks and sometimes-finer disks are used for improving the surface quality. Figure 4.15 shows a cross sectional view of a sample where the gray areas are the fiber/matrix and the dark areas are those corresponding to porosity. The image was processed through an m-code

Table 4.3 NDT standards with further information.

Organi-zation	Standard number	Title
ASTM	E2862	Standard Practice for Probability of Detection Analysis for Hit/Miss Data
ASTM	E1067	Standard Practice for Acoustic Emission Examination of Fiberglass Reinforced Plastic Resin (FRP) Tanks/Vessels
ASTM	E1118	Standard Practice for Acoustic Emission Examination of Reinforced Thermosetting Resin Pipe (RTRP)
ASTM	E1495	Standard Guide for Acousto-Ultrasonic Assessment of Composites, Laminates, and Bonded Joints
ASTM	E2076	Examination of Fiberglass Reinforced Plastic Fan Blades Using Acoustic Emission
ASTM	E2191	Standard Practice for Examination of Gas-Filled Filament-Wound Composite Pressure Vessels Using Acoustic Emission
ASTM	E2533	Standard Guide for Nondestructive Testing of Polymer Matrix Composites Used in Aerospace Applications
ASTM	E2580	Standard Practice for Ultrasonic Testing of Flat Panel Composites and Sandwich Core Materials Used in Aerospace Applications
ASTM	E2581	Standard Practice for Shearography of Polymer Matrix Composites and Sandwich Core Materials in Aerospace Applications
ASTM	E2582	Standard Practice for Infrared Flash Thermography of Composite Panels and Repair Patches Used in Aerospace Applications
ASTM	E2661	Standard Practice for Acoustic Emission Examination of Plate-like and Flat Panel Composite Structures Used in Aerospace Applications
ASTM	E2662	Standard Practice for Radiographic Examination of Flat Panel Composites and Sandwich Core Materials Used in Aerospace Applications
ASTM	E2981	Standard Guide for Nondestructive Testing of the Composite Overwraps in Filament Wound Pressure Vessels Used in Aerospace Applications
BS	BS EN ISO 11623	Gas cylinders. Composite construction. Periodic inspection and testing
BS	BS EN-1330	Terms Used in Acoustic Emission Testing.
BS	BS EN-13477	Acoustic Emission Testing
BS	BS EN-25580	Specification for Minimum Requirements for Industrial Radiographic Illuminators for NDT
ISO	ISO 18249	Non-destructive testing – Acoustic emission testing – Specific methodology and general evaluation criteria for testing of fiber-reinforced polymers
ISO	ISO 3058	Non-destructive testing – Aids to visual inspection – Selection of low-power magnifiers

ASTM – American Society for Testing and Materials, BS – British Standards, ISO – International Organization for Standardization.

program using the grayscale histogram method (Matlab; Mathworks, Natick, MA, USA) [30]. The pixels of the image are plotted according its intensity and frequency of occurrence. The porosity level is determined according to the ratio of the sum of the dark pixels identified to the total number of pixels. This method is only valid for the scans at the slice analyzed and may not be representative of porosity in other regions.

4.9   NDT Standards

Table 4.3 shows NDT standards related to the contents of this chapter. Standards are continuously updated, thus it is advisable to check for the most recent version.

References

1 Good, A. and Steiner, D. (2017). X-rays for NDT of composites. In: CompositesWorld. Gardner Business Media: Cincinnati, OH.

2 Yang, L. and Hung, Y. (2004) Digital shearography for nondestructive evaluation and application in automotive and aerospace industries. Journal of Holography and Speckle 1 (2): 69–79.

3 Cemenska, J., Rudberg, T., and Henscheid, M. (2015). Automated in-process inspection system for AFP machines. SAE International Journal of Aerospace 8 (2015-01-2608): 303–309.

4 BDM Federal, I. (1998). Corrosion Detection Technologies, Sector Study. Prepared for the North American Technology and Industrial Base Organization (NATIBO).

5 Dance, W. (1976). Neutron radiographic nondestructive evaluation of aerospace structures. In: Practical Applications of Neutron Radiography and Gaging. ASTM International.

6 Kessler, S.S., Spearing, S.M., and Soutis, C. (2002). Damage detection in composite materials using Lamb wave methods. Smart Materials and Structures 11 (2): 269.

7 Hamstad, M.A. (1986). A review: acoustic emission, a tool for composite-materials studies. Experimental Mechanics 26 (1): 7–13.

8 Cawley, P. and Theodorakopoulos, C. (1989). The membrane resonance method of non-destructive testing. Journal of sound and vibration 130 (2): 299–311.

9 Steinchen, W. and Yang, L. (2003). Digital Shearography: Theory and Application of Digital Speckle Pattern Shearing Interferometry. Bellingham: Spie Press.

10 Klute, S. (2012). Effects of defects, in Aerospace Testing International. Surrey, UK.

11 Adams, R. and Cawley, P. (1988). A review of defect types and non-destructive testing techniques for composites and bonded joints. NDT International 21 (4): 208–222.

12 El Guerjouma, R., Baboux, J.-C., Ducret, D. et al. (2001). Non-destructive evaluation of damage and failure of fibre reinforced polymer composites using ultrasonic waves and acoustic emission. Advanced Engineering Materials 3 (8): 601–608.

13 Qamhia, I., Lauer-Hunt, E., and Elhajjar, R. (2013). Identification of acoustic emissions from porosity and waviness defects in continuous fiber reinforced composites. Advances in Civil Engineering Materials 2 (1): 37–50.

14 Chakrapani, S.K., Barnard, D., and Dayal, V. (2014). Detection of in-plane fiber waviness in composite laminates using guided Lamb modes. In: 40th Annual Review of Progress in Quantitative Nondestructive Evaluation: Incorporating the 10th International Conference on Barkhausen Noise and Micromagnetic Testing. AIP Publishing.

15 Chakrapani, S.K., Dayal, V., Hsu, D.K., Barnard, D.J., Gross, A., Thompson, D.O., and Chimenti, D.E. Characterization of waviness in wind turbine blades using air coupled ultrasonics. in AIP Conference Proceedings. 2011. AIP.

16 Wooh, S.C. and Daniel, I.M. (1995). Wave-propagation in composite-materials with fiber waviness. Ultrasonics 33 (1): 3–10.

17 Nikishkov, Y., Airoldi, L., and Makeev, A. (2013). Measurement of voids in composites by X-ray computed tomography. Composites Science and Technology 89: 89–97.

18 10-0369-EL-BGN, C.N.(2012). Correspondence from EDP Renewables North America LLC electronically filed by Mr. Michael J. Settineri on Behalf of EDP Renewables North America LLC. Columbus, OH: The Public Utilities Commission of Ohio.

19 Elhajjar, R.F. and Shams, S.S. (2016). A new method for limit point determination in composite materials containing defects using image correlation. Composites Science and Technology 122: 140–148.

20 Henry, T.C., Bakis, C.E., Riddick, J.C., and Smith, E.C. (2014). Full-field strain analysis of compressively loaded flat composite laminates with undulated fibers. In: Proceedings of the American Society for Composites 2014-Twenty-Ninth Technical Conference on Composite Materials. DEStech Publications, Inc.

21 Elhajjar, R.F., Shams, S.S., Kemeny, G.J., and Stuessy, G. (2016). A hybrid numerical and imaging approach for characterizing defects in composite structures. Composites Part A: Applied Science and Manufacturing 81: 98–104.

22 Shelley, P.H., Werner, G.J., Vahey, P.G., Kemeny, G.J., Groth, G.A., Stuessy, G., and Zueger, C., Near infrared imaging for measuring resin thickness on composites, in SAMPE Annual Conference. 2014, Society for the Advancement of Material and Process Engineering: Seattle, WA.

23 Elhajjar, R., Haj-Ali, R., and Wei, B.-S. (2014). An infrared thermoelastic stress analysis investigation for detecting fiber waviness in composite structures. Polymer-Plastics Technology and Engineering 53 (12): 1251–1258.

24 Sullivan, R.W., Balasubramaniam, K., and Bennett, G. (1996). Plate wave flow patterns for ply orientation imaging in fiber reinforced composites. Materials Evaluation 54 (4): 518–523.

25 Maffezzoli, A., Quarta, E., Luprano, V. et al. (1999). Cure monitoring of epoxy matrices for composites by ultrasonic wave propagation. Journal of Applied Polymer Science 73 (10): 1969–1977.

26 Shepard, D.D. and Smith, K.R. (1999). Ultrasonic cure monitoring of advanced composites. Sensor Review 19 (3): 187–194.

27 Antonucci, V., Giordano, M., Cusano, A. et al. (2006). Real time monitoring of cure and gelification of a thermoset matrix. Composites Science and Technology 66 (16): 3273–3280.

28 Schubel, P., Crossley, R., Boateng, E., and Hutchinson, J. (2013). Review of structural health and cure monitoring techniques for large wind turbine blades. Renewable Energy 51: 113–123.

29 Hsu, D.K. and Uhl, K.M. (1987). A morphological study of porosity defects in graphite-epoxy composites. In: Review of Progress in Quantitative Nondestructive Evaluation, 1175–1184. Springer.

30 Yang, P. and El-Hajjar, R. (2012). Porosity defect morphology effects in carbon fiber–epoxy Composites. Polymer-Plastics Technology and Engineering 51 (11): 1141–1148.

31 Kastner J, Plank B., Salaberger D, Sekelja J. Defect and porosity determination of fibre reinforced polymers by x-ray computed tomography. in 2nd International Symposium on NDT in Aerospace 2010. 2010.

32 Hungler, P., Bennett, L., Lewis, W. et al. (2011). Neutron imaging inspections of composite honeycomb adhesive bonds. Nuclear Instruments and Methods in Physics Research Section A: Accelerators, Spectrometers, Detectors and Associated Equipment 651 (1): 250–252.

33 Hodge, A. and Dambaugh, G. (2013). Analysis of thermally induced stresses on the core node bonds of a co-cured sandwich panel. Journal of Composite Materials 47 (4): 467–474.

34 Kim, K.H., Klann, R.T., and Raju, B.B. (1999). Fast neutron radiography for composite materials evaluation and testing. Nuclear Instruments and

Methods in Physics Research Section A: Accelerators, Spectrometers, Detectors and Associated Equipment 422 (1): 929–932.

35 Mustapha, S., Ye, L., Dong, X., and Alamdari, M.M. (2016). Evaluation of barely visible indentation damage (BVID) in CF/EP sandwich composites using guided wave signals. Mechanical Systems and Signal Processing 76: 497–517.

36 Cawley, P. (1987). The sensitivity of the mechanical impedance method of nondestructive testing. NDT International 20 (4): 209–215.

37 Esquivel, O. and Seibold, R.W. (2004). Capabilities and Limitations of Nondestructive Evaluation Methods for Inspecting Components Beneath Thermal Protection Systems, DOT-VNTSC-FAA-04-10. El Segundo, CA 90245-4691: The Aerospace Corporation.

38 Guyott, C. and Cawley, P. (1988). Evaluation of the cohesive properties of adhesive joints using ultrasonic spectroscopy. NDT International 21 (4): 233–240.

39 Nagy, P.B. (1992). Ultrasonic classification of imperfect interfaces. Journal of Nondestructive evaluation 11 (3–4): 127–139.

40 Wood, M., Charlton, P., and Yan, D. (2014) Ultrasonic evaluation of artificial kissing bonds in CFRP composites. The e-Journal of NDT.

41 Adams, R. and Drinkwater, B. (1997). Nondestructive testing of adhesively-bonded joints. NDT & E International 30 (2): 93–98.

42 Cawley, P. and Adams, R. (1988). The mechanics of the coin-tap method of non-destructive testing. Journal of Sound and Vibration 122 (2): 299–316.

43 Georgeson, G.E., Lea, S., and Hansen, J. (1996). Electronic tap hammer for composite damage assessment. In: Nondestructive Evaluation Techniques for Aging Infrastructure and Manufacturing. International Society for Optics and Photonics.

44 Cawley, P. (1984). The impedance method of non-destructive inspection. NDT International 17 (2): 59–65.

45 Hung, Y., Luo, W., Lin, L., and Shang, H. (2000). Evaluating the soundness of bonding using shearography. Composite Structures 50 (4): 353–362.

46 Pavlopoulou, S., Soutis, C., and Manson, G. (2012). Non-destructive inspection of adhesively bonded patch repairs using Lamb waves. Plastics, Rubber and Composites 41 (2): 61–68.

47 Poudel, A. (2015). Bond Strength Evaluation in Adhesive Joints Using NDE and DIC Methods. Southern Illinois University at Carbondale: ProQuest Dissertations Publishing. 3716032.

48 Kumar, R.V., Bhat, M., and Murthy, C. (2013). Evaluation of kissing bond in composite adhesive lap joints using digital image correlation: preliminary studies. International Journal of Adhesion and Adhesives 42: 60–68.

49 Liu, Y., Johnston, J., and Chattopadhyay, A. (2013). Non-destructive evaluation of composite adhesive kissing bond. In: ASME 2013 International Mechanical Engineering Congress and Exposition. American Society of Mechanical Engineers.

50 Seaton, C. and Richter, S. DOT/FAA/TC-14/20(2014). Nonconforming Composite Repairs: Case Study Analysis. Federal Aviation Administration, William J. Hughes Technical Center Aviation Research Division Atlantic City International Airport New Jersey 08405.

51 Stone, D. and Clarke, B. (1975). Ultrasonic attenuation as a measure of void content in carbon-fibre reinforced plastics. Non-Destructive Testing 8 (3): 137–145.

52 Smith, R.A., Nelson, L.J., Mienczakowski, M.J., and Challis, R.E. (2009). Automated analysis and advanced defect characterisation from ultrasonic scans of composites. Insight-Non-Destructive Testing and Condition Monitoring 51 (2): 82–87.

53 Jeong, H. and Hsu, D.K. (1995). Experimental analysis of porosity-induced ultrasonic attenuation and velocity change in carbon composites. Ultrasonics 33 (3): 195–203.

54 Daniel, I., Wooh, S., and Komsky, I. (1992). Quantitative porosity characterization of composite materials by means of ultrasonic attenuation measurements. Journal of Nondestructive Evaluation 11 (1): 1–8.

55 ASTM, E2862/E2862 - 12(2012). Standard Practice for Probability of Detection Analysis for Hit/Miss Data. West Conshohocken, PA: ASTM International.

56 1823A, M.-H., Nondestructive Evaluation System Reliability Assessment. Standardization Order Desk, Building 4D, 700 Roberts Avenue, Philadelphia (2009). PA. In: 2009.

57 Barnard, D., Foreman, C., and Hsu, D. Generating tap test images with a free-hand motorized tapper. In Review of Progress in Quantitative Nondestructive Evaluation: Proceedings of the 35th Annual Review of Progress in Quantitative Nondestructive Evaluation. 2009. AIP Publishing. .

58 Hsu, D., Barnard, D., Peters, J., and Dayal, V. Development of nondestructive inspection methods for composite repair. in AIP Conference Proceedings. 2003. IOP Institute of Physics Publishing Ltd.

59 Barnard, D.J., Peters, J.J., and Hsu, D.K. (2001). Development of a magnetic cam for the computer aided tap test system. In: Review of Progress in Quantitative Nondestructive Evaluation, vol. 20. AIP Publishing.

60 Frankle, R.S. and Rose, D.N. (1995). Flexible ultrasonic array system for inspecting thick composite structures. In: Nondestructive Evaluation of Aging Infrastructure. International Society for Optics and Photonics.

61 Baid, H., Schaal, C., Samajder, H., and Mal, A. (2015). Dispersion of Lamb waves in a honeycomb composite sandwich panel. Ultrasonics 56: 409–416.

62 Weber, R., Hosseini, S.M.H., and Gabbert, U. (2012). Numerical simulation of the guided Lamb wave propagation in particle reinforced composites. Composite Structures 94 (10): 3064–3071.

63 Leleux, A., Micheau, P., and Castaings, M. (2013). Long range detection of defects in composite plates using Lamb waves generated and detected by ultrasonic phased array probes. Journal of Nondestructive Evaluation 32 (2): 200–214.

64 Li, W., Cho, Y., and Achenbach, J.D. (2012). Detection of thermal fatigue in composites by second harmonic Lamb waves. Smart Materials and Structures 21 (8): 085019.

65 Tian, Z., Yu, L., and Leckey, C. (2015). Delamination detection and quantification on laminated composite structures with Lamb waves and wavenumber analysis. Journal of Intelligent Material Systems and Structures 26 (13): 1723–1738.

66 Sreekumar, P., Ramadas, C., Anand, A., and Joshi, M. (2015). Attenuation of Ao Lamb mode in hybrid structural composites with nanofillers. Composite Structures 132: 198–204.

67 Torkamani, S., Roy, S., Barkey, M.E. et al. (2014). A novel damage index for damage identification using guided waves with application in laminated composites. Smart Materials and Structures 23 (9): 095015.

68 Habib, F., Martinez, M., Artemev, A., and Brothers, M. (2013). Structural health monitoring of bonded composite repairs–a critical comparison between ultrasonic Lamb wave approach and surface mounted crack sensor approach. Composites Part B: Engineering 47: 26–34.

69 Sherafat, M.H., Quaegebeur, N., Lessard, L., Hubert, P., and Masson, P. Guided wave propagation through composite bonded joints. in EWSHM-7th European Workshop on Structural Health Monitoring. 2014.

70 Su, Z., Ye, L., and Lu, Y. (2006). Guided Lamb waves for identification of damage in composite structures: a review. Journal of Sound and Vibration 295 (3–5): 753–780.

71 Viktorov, I.A. (1967). Rayleigh and Lamb Waves: Physical Theory and Applications, vol. 147. New York: Plenum press.

72 Schmidt, F., Rheinfurth, M., Horst, P., and Busse, G. (2012). Effects of local fibre waviness on damage mechanisms and fatigue behaviour of biaxially loaded tube specimens. Composites Science and Technology 72 (10): 1075–1082.

73 Williams, R.V. (1980). Acoustic Emission, 118. Bristol: Adam Hilger Ltd.

74 Williams, J.H. and Lee, S.S. (1978). Acoustic emission monitoring of fiber composite materials and structures. Journal of Composite Materials 12 (4): 348–370.

75 Chang, R. (2000). Experimental and theoretical analyses of first-ply failure of laminated composite pressure vessels. Composite Structures 49 (2): 237–243.

76 Walker, J.L., Workman, G.L., Russell, S.S., and Hill, E. (1997). Neural network/acoustic emission burst pressure prediction for impact damaged composite pressure vessels. Materials Evaluation 55 (8).

77 Hill, E.V. (1992). Predicting burst pressures in filament-wound composite pressure vessels by using acoustic emission data. Materials Evaluation 50: 1439–1445.

78 Fowler, T., The Origin of CARP and the Term "Felicity Effect", in 31st Conference of the European Working Group on Acoustic Emission (EWGAE) – Commemorative Speech. 2014: Dresden, Germany.

79 Jemielniak, K. (2001). Some aspects of acoustic emission signal pre-processing. Journal of Materials Processing Technology 109 (3): 242–247.

80 Promboon, Y. (2000). Acoustic Emission Source Location, 343. The University of Texas at Austin.

81 Ziola, S.M. and Gorman, M.R. (1991). Source location in thin plates using cross-correlation. Journal of the Acoustical Society of America 90 (5): 2551–2556.

82 Dzenis, Y.A. and Saunders, I. (2002). On the possibility of discrimination of mixed mode fatigue fracture mechanisms in adhesive composite joints by advanced acoustic emission analysis. International Journal of Fracture 117 (4): L23–L28.

83 Godin, N., Momon, S., Moevus, M. et al. (2010). Acoustic emission and lifetime prediction during static fatigue tests on ceramic-matrix-composite at high temperature under air. Composites Part A Applied Science and Manufacturing 41 (7): 913–918.

84 Biermann, H., Vinogradov, A., and Hartmann, O. (2002). Fatigue damage evolution in a particulate-reinforced metal matrix composite determined by acoustic emission and compliance method. Zeitschrift Fur Metallkunde 93 (7): 719–723.

85 Schiavon, I., Fleischmann, P., Fougeres, R., and Rouby, D. (1987). Fatigue damage of one-dimensional glass epoxy composite controlled by mechanical parameters and acoustic-emission. Journal of Materials Science Letters 6 (10): 1182–1184.

86 Awerbuch, J. and Ghaffari, S. (1988). Monitoring progression of matrix splitting during fatigue loading through acoustic-emission in notched unidirectional graphite epoxy composite. Journal of Reinforced Plastics and Composites 7 (3): 245–264.

87 Roy, C. and Elghorba, M. (1988). Monitoring progression of mode-ii delamination during fatigue loading through acoustic-emission in laminated glass-fiber composite. Polymer Composites 9 (5): 345–351.

88 Gilblom, D.L., Colbeth, R.E., Batts, M., and Meyer, B. Real-time X-ray imaging with flat panels. in Non-Destructive Evaluation Techniques for

Aging Infrastructure & Manufacturing. 1998. International Society for Optics and Photonics.

89 Miracle, D.B., Donaldson, S.L., Henry, S.D. et al. (2001). ASM Handbook, vol. 21. Materials Park, OH: ASM International.

90 Kortschot, M. and Zhang, C. (1995). Characterization of composite mesostructures and damage by de-ply radiography. Composites Science and Technology 53 (2): 175–181.

91 Yang, P. and Elhajjar, R. (2014). Porosity content evaluation in carbon-fiber/epoxy composites using X-ray computed tomography. Polymer-Plastics Technology and Engineering 53 (3): 217–222.

92 Flavio De Andrade Silva, J.W., Muller, B.R., Hentschel, M.P. et al. (2010). Three-dimensional microstructure visualization of porosity and Fe-rich inclusions in SiC particle-reinforced Al ally matrix composites by X-ray synchrotron tomography. Metallurgical and Materials Transaction A 41 (8): 2121–2128.

93 Williams, J.J., Flom, Z., Amell, A.A. et al. (2010). Damage evolution in SiC particle reinforced Al alloy matrix composites by X-ray synchrotron tomography. Acta Materialia 58: 6194–6205.

94 Schilling, P.J., Karedla, B.P.R., Tatiparthi, A.K. et al. (2005). X-ray computed microtomography of internal damage in fiber reinforced polymer matrix composites. Composites Science and Technology 65: 2071–2078.

95 Write, P., Fu, X., Sinclair, I., and Spearing, S.M. (2008). Ultra-high resolution computed tomography of damage in notched carbon fiber-epoxy composites. Journal of Composite Materials 42: 1993–2002.

96 McMillan, A.J. (2012). Material strength knock-down resulting from multiple randomly positioned voids. Journal of Reinforced Plastics and Composites 31 (1): 13–28.

97 Harding, G. and Kosanetzky, J. (1989). Scattered X-ray beam nondestructive testing. Nuclear Instruments and Methods in Physics Research Section A: Accelerators, Spectrometers, Detectors and Associated Equipment 280 (2–3): 517–528.

98 Hung, Y. (1996). Shearography for non-destructive evaluation of composite structures. Optics and Lasers in Engineering 24 (2): 161–182.

99 Hung, Y. (1999). Applications of digital shearography for testing of composite structures. Composites Part B: Engineering 30 (7): 765–773.

100 Santos, F., Vaz, M., and Monteiro, J. (2004). A new set-up for pulsed digital shearography applied to defect detection in composite structures. Optics and Lasers in Engineering 42 (2): 131–140.

101 Toh, S., Shang, H., Chau, F., and Tay, C. (1991). Flaw detection in composites using time-average shearography. Optics & Laser Technology 23 (1): 25–30.

102 Garnier, C., Pastor, M.-L., Eyma, F., and Lorrain, B. (2011). The detection of aeronautical defects in situ on composite structures using non destructive testing. Composite Structures 93 (5): 1328–1336.

103 De Angelis, G., Meo, M., Almond, D.P. et al. (2012). A new technique to detect defect size and depth in composite structures using digital shearography and unconstrained optimization. NDT & E International 45 (1): 91–96.

104 Toh, S., Chau, F., Shim, V. et al. (1990). Application of shearography in nondestructive testing of composite plates. Journal of Materials Processing Technology 23 (3): 267–275.

105 Lee, J.-R., Molimard, J., Vautrin, A., and Surrel, Y. (2004). Digital phase-shifting grating shearography for experimental analysis of fabric composites under tension. Composites Part A: Applied Science and Manufacturing 35 (7): 849–859.

106 Ryley, A.C., Kharkovsky, S., Daniels, D., Kreitinger, N., Steffes, G., Zoughi, R., and Abou-Khousa, M.A., Comparison of X-ray, millimeter wave, shearography and through-transmission ultrasonic methods for inspection of honeycomb composites. 2007.

107 Goidescu, C., Welemane, H., Garnier, C. et al. (2013). Damage investigation in CFRP composites using full-field measurement techniques: combination of digital image stereo-correlation, infrared thermography and X-ray tomography. Composites Part B: Engineering 48: 95–105.

108 Johanson, K., Harper, L.T., Johnson, M.S., and Warrior, N.A. (2015). Heterogeneity of discontinuous carbon fibre composites: damage initiation captured by digital image correlation. Composites Part A: Applied Science and Manufacturing 68: 304–312.

109 Makeev, A., Seon, G., and Lee, E. (2009). Failure predictions for carbon/epoxy tape laminates with wavy plies. Journal of Composite Materials.

110 El-Hajjar, R.F. and Petersen, D.R. (2011). Gaussian function characterization of unnotched tension behavior in a carbon/epoxy composite containing localized fiber waviness. Composite Structures 93 (9): 2400–2408.

111 Lang, E.J. and Chou, T.-W. (1998). The effect of strain gage size on measurement errors in textile composite materials. Composites Science and Technology 58 (3): 539–548.

112 Shukla, A., Agarwal, B., and Bhushan, B. (1989). Determination of stress intensity factor in orthotropic composite materials using strain gages. Engineering Fracture Mechanics 32 (3): 469–477.

113 di Scalea, F.L. (1998). Measurement of thermal expansion coefficients of composites using strain gages. Experimental Mechanics 38 (4): 233–241.

114 Murukeshan, V., Chan, P., Ong, L., and Seah, L. (2000). Cure monitoring of smart composites using fiber Bragg grating based embedded sensors. Sensors and Actuators A: Physical 79 (2): 153–161.

115 Childers, B.A., Froggatt, M.E., Allison, S.G. et al. (2001). Use of 3000 Bragg grating strain sensors distributed on four 8-m optical fibers during static load tests of a composite structure. In: SPIE's 8th Annual International Symposium on Smart Structures and Materials. International Society for Optics and Photonics.

116 Watkins, S.E., Sanders, G.W., Akhavan, F., and Chandrashekhara, K. (2002). Modal analysis using fiber optic sensors and neural networks for prediction of composite beam delamination. Smart Materials and Structures 11 (4): 489.

117 Baldwin, C.S., Poloso, T., Chen, P.C. et al. (2001). Structural monitoring of composite marine piles using fiber optic sensors. In: SPIE's 8th Annual International Symposium on Smart Structures and Materials. International Society for Optics and Photonics.

118 Maaskant, R., Alavie, T., Measures, R. et al. (1997). Fiber-optic Bragg grating sensors for bridge monitoring. Cement and Concrete Composites 19 (1): 21–33.

119 Kister, G., Winter, D., Badcock, R. et al. (2007). Structural health monitoring of a composite bridge using Bragg grating sensors. Part 1: evaluation of adhesives and protection systems for the optical sensors. Engineering Structures 29 (3): 440–448.

120 Seim, J., Udd, E., Schulz, W., and Laylor, H.M. Composite strengthening and instrumentation of the Horsetail Falls Bridge with long gage length fiber Bragg grating strain sensors. In Proceedings-Spie the International Society for Optical Engineering. 1999. Spie International Society for Optical. .

121 Bocherens, E., Bourasseau, S., Dewynter-Marty, V. et al. (2000). Damage detection in a radome sandwich material with embedded fiber optic sensors. Smart Materials and Structures 9 (3): 310.

122 Claus, R.O., Holton, C.E., and Zhao, W. (1998). Performance of optical fiber sensors embedded in polymer matrix composites for 15 years. In: 5th Annual International Symposium on Smart Structures and Materials. International Society for Optics and Photonics.

123 Schroeder, J., Ahmed, T., Chaudhry, B., and Shepard, S. (2002). Non-destructive testing of structural composites and adhesively bonded composite joints: pulsed thermography. Composites Part A: Applied Science and Manufacturing 33 (11): 1511–1517.

124 Genest, M., Martinez, M., Mrad, N. et al. (2009). Pulsed thermography for non-destructive evaluation and damage growth monitoring of bonded repairs. Composite Structures 88 (1): 112–120.

125 Sun, J. (2006). Analysis of pulsed thermography methods for defect depth prediction. Journal of Heat Transfer 128 (4): 329–338.

126 Benítez, H.D., Ibarra-Castanedo, C., Bendada, A. et al. (2008). Definition of a new thermal contrast and pulse correction for defect quantification in pulsed thermography. Infrared Physics & Technology 51 (3): 160–167.

127 Avdelidis, N.P., Almond, D.P., Dobbinson, A. et al. (2004). Aircraft composites assessment by means of transient thermal NDT. Progress in Aerospace Sciences 40 (3): 143–162.

128 Avdelidis, N., Hawtin, B., and Almond, D. (2003). Transient thermography in the assessment of defects of aircraft composites. Ndt & E International 36 (6): 433–439.

129 Anastasi, R.F., Zalameda, J.N., and Madaras, E.I., Damage Detection in Rotorcraft Composite Structures Using Thermography and Laser-Based Ultrasound. 2004.

130 Hayes, B.S. and Gammon, L.M. (2010). Optical Microscopy of Fiber-Reinforced Composites. ASM international.

131 ASTM, D 2734(2016). Standard Test Methods for Void Content of Reinforced Plastics. West Conshohocken, PA: ASTM International.

132 ASTM, D3171(2015). Standard Test Methods for Constituent Content of Composite Materials. West Conshohocken, PA: ASTM International.

133 Yang, P., Qamhia, I., Shams, S., and Elhajjar, R., Dynamic Mechanical Characterization of Manufacturing Defects in Continuous Carbon-Fiber/Epoxy Composites. SAMPE Conference Proceedings. Seattle, WA. Society for the Advancement of Material and Process Engineering., 2014.

5

Effects of Defects – Design Values and Statistical Considerations

5.1 Introduction

In this chapter, we present methods that can be used to evaluate the effects of common defects analytically and experimentally. The chapter includes references to some of the most significant studies in this area that may serve as guidance for further analysis or experimentation. Despite the prevalence of many studies on the effects of defects in composite materials, the area is still largely ripe for further study. The diversity of fiber and matrix material choices, fiber treatments, and possible laminate configurations makes the cost of a large-scale experimental effort to study the strength reductions of defects a complex task. Further, test coupons used to produce effective strength reductions do not necessarily replicate the multiaxial stress state that is present in real structures due to the unique notched behavior for each laminate configuration and damage state. The failure characteristics of structural components are generally more complex than those of the relatively simple coupon specimens often used for material characterization and the development of structural design values, primarily due to the complex in-plane (often biaxial) and out-of-plane loading modes present in real structures. Real structures and large components may also respond with changes in the overall structure, as in during buckling, for example. When considering effects of defects on materials it is also important to note that load redistribution may occur in the real structure, which may not occur in the simple tests. This load redistribution may manifest itself in different types of stress, such as large interlaminar stresses in the component. Thus, extrapolating from coupon data to large-scale structural components has its limitations. In a study performed by Marrouze et al. [1], finite-element computational approaches with micromechanics were used to investigate reductions in behavior in stiffened compression panels with gaps and fiber waviness. They found from the structural models that, because of the load redistribution, there were lower strength and stiffness reductions than those predicted from coupon tests. A robust building block leading to validated analysis addressed

Composite Structures: Effects of Defects, First Edition.
Rani Elhajjar, Peter Grant and Cindy Ashforth.
© 2019 John Wiley & Sons Ltd. Published 2019 by John Wiley & Sons Ltd.

Table 5.1 Suggested combined loadings to consider for in-plane panel tests.

Load conditions	
Maximum tension	In-plane shear
Maximum compression	In-plane shear
Maximum in-plane shear	Compression
Maximum in-plane shear	Tension

these challenges. In lieu of a complex test and analytical database, simplified methods may be used when shown to be conservative.

Approaches used to determine the effect of a particular defect upon structural strength and/or stiffness may vary depending upon the defect size. The smaller defects may be evaluated at the coupon level or possibly up to the structural element level (see the building block chart in Figure 2.1). Coupon-level tests ignore the effect of load redistribution, which in larger structures may result in an increased level of conservatism. However, the coupon testing approach is generally much more economical. Larger damage sizes may have to be evaluated in a more complex test item (with usually one replicate), normally involving a repaired or damaged structure. If larger panels are considered, it is important to consider the different loading envelopes as shown in Table 5.1. The tests may have to be repeated for limit and ultimate conditions.

Retention factors are determined from a testing program on coupons or small elements manufactured with expected defects and typical laminates. The retention or knockdown factor is the ratio of the mechanical property accounting for the flaw to the mechanical property in the unflawed condition. The values can range from zero to one, with a value of one indicating no effect from the defect. *Strength retention factors* are the strength of the baseline specimen incorporating a defect divided by the strength of a baseline coupon or element as used in the development of design values. The baseline coupons usually have a filled or open hole. Retention factors for notched strength, stiffness, and fracture toughness properties will be necessary depending on the analysis performed.

It is the responsibility of the structural analyst to ensure that they have captured the manufacturing anomaly effects on all the variables that may affect the results of their structural analysis. This task should also consider the possibility of multiple defects occurring at the same time. It is generally not advisable to perform effects of defect evaluations on quality control type tests which may have other intentional damages in the test article configuration (e.g., compression after impact). Such tests may be used for screening purposes but their direct use in design has not been established. Thus, it is generally advisable that effects-of-defects testing be performed on standard test specimens used to obtain *design allowables*. Use of non-standard tests can result in misleading results that can be over-conservative in many cases and not capture the true impact of the defect.

In general, for *element level* testing, the width of the defect should not exceed 15% of the width of the panel and defects should not be permitted within a zone of 2.5 times the thickness from any edge. A 1 cm/0.5 in. flaw is typically included in analysis and design purposes for disbonds (e.g., face to core). Minimum separation distances for multiple defects should also be considered. For example, critical structures may require larger spaces than other structures (Figure 5.1).

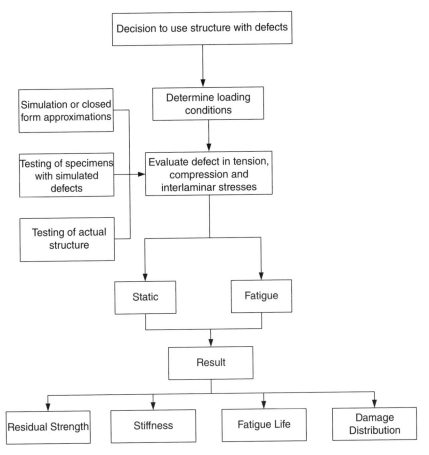

Figure 5.1 Steps in decision making when evaluating a structure with defects.

5.2 Effects on Laminate Properties

5.2.1 Cure Cycle Anomalies

During the processing of composite structures, it is not uncommon to encounter cure cycle anomalies. Regulatory agencies such as the Federal Aviation Administration (FAA) will typically require the manufacturer to show that they have substantiated their structures with the entire envelope of processing parameters allowed by their design/specifications. Sophisticated methods can be used to model processing parameters, find the most critical ones, and test those. Simplistic methods are to limit the number and scope of cure cycles to just one or two, and approve the processing envelope through testing the various configurations during certification.

With the analytical method, one may be able to approve processing deviations analytically if they fall within the limits of the analysis (i.e., it is okay to interpolate, but not extrapolate). Otherwise testing is typically used to disposition the deviation. It could be a series of test panels run at that cure cycle, testing on a traveler coupon, or it could be sacrificing a part in the oven run.

The manufacturer could also set a smaller processing window, when they have already approved much larger limits, but do not use all the tolerance in initial design. The benefits of such an approach is that some margin is kept in reserve for deviations that are certain to occur eventually. The following form list of some of the cure cycle anomalies that can be encountered during the curing of thermoset-based composites,

- Rapid or slow ramp rate of temperature to first dwell
- Exotherm during first or second ramp
- Extended dwell time after ramp
- Shortened dwell after ramp
- Extended dwell time at maximum temperature
- Loss of vacuum during certain stages

5.2.1.1 Porosity

This topic, as presented in Chapter 3, has been extensively examined in the literature and the typical retention factors reported tend to be valid if the porosity is distributed through the cross-section [2–12]. Porosity is defined as a small cluster of voids internal to the cured laminate. A cluster of voids shall be considered porosity if the maximum dimension of the largest void is less than approximately 5 mm. Typically, there is a minimal effect on the mechanical properties of porosity up to 2% and frequently the effects of porosity at this level are already included in the design values. There is a direct correlation between high porosity and several mechanical properties. Porosity has a large effect on resin-dominated mechanical properties, such as compressive strength and interlaminar shear strength. Out-of-plane shear stiffness is also significantly impacted as evidenced by direct correlation between short-beam shear strength results and porosity content.

Porosity does not have a large effect on fiber-dominated mechanical properties, such as tensile strength in the fiber direction. A higher bending stiffness in a laminate maybe realized in some cases due to increased thickness associated with porosity. Accurate quantification on the nature of porosity above 8% using immersion UT (ultrasonic transmission) is difficult and there is limited evidence to support structures operated outside this range. There are also many mature methods for nondestructive testing that can be used within the limits described previously. The typical retention values do not apply to large clusters of coalesced voids. In these cases, these formations may be considered to be delaminations and examined accordingly.

5.2.1.2 Nonuniformly Distributed Voids (Stratified Porosity)

In some rare cases of processing, incoming material, and quality-induced errors can result in nonuniformly distributed voids (through-the-thickness) in the composite laminates. Standard methods (e.g., acid digestion) may indicate a low or average void content, but that can mask the through-the-thickness nonuniformity. This type of defect may be caused by a low quality region in the incoming material, poor layup technique resulting in entrapped air or an incorrect synchronization of pressure and temperature during the curing process. One study has shown mild strength reductions except for elevated temperatures where at 265 °F, the non-uniform void specimens achieved less than 25% of the compression strength compared to their void-free counterparts [13]. Specimens with uniform void content were able to achieve approximately 80% of the compression strength at that temperature.

5.2.1.3 Stratified Porosity or Delamination in L-Shaped Details

The presence of porosity in radius details can reduce the flange bending strength and overall of the structure due to initiation of delaminations and their growth. The physics behind this process is shown in Figure 5.2 [14]. In this analysis we can see that point 1 is where the predicted delamination emergence load is reached. Applying additional force leads to a change of the structural stiffness and to a reduction of the reaction force at point 2. Beyond point 2, still maintaining the same displacement load, unstable delamination growth takes place, which leads to a considerable increase of the delaminated area (point 3) followed by stable crack growth until point 4. Unloading and reloading a structure after the delamination has grown can be expected to lead to a significantly reduced stiffness and strength. Similarly, manufacturing-induced stratified porosity and delamination at the radius area can result in a condition where the stiffness and strength are significantly reduced.

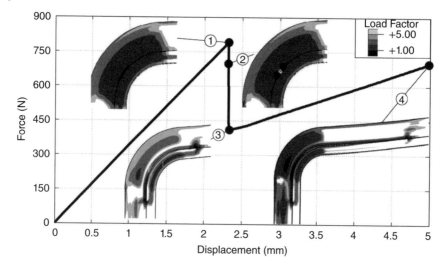

Figure 5.2 Delamination process in the radius detail of an L-shaped laminate. Source: Reproduced with permission from Elsevier.

5.2.1.4 In-Plane Fiber Misalignment

Fiber misalignment can occur due to changes in geometry and curvature and can contribute to imbalance or asymmetry in the laminate. Evaluations performed on graphite/epoxy unidirectional laminates using a micromechanical analysis that account for the associated local field distributions within the constituent material have been performed [15]. In this approach, the probability-weighted averaging of the appropriate concentration tensors, are used to determine damage initiation envelopes, local fields, effective moduli, and strengths of the unidirectional composite behavior. This approach can be used to study the effect on allowables of the statistical information on the misalignment is available. In wind turbine blades, a 1–5% misalignment (imbalance) can occur in the turbine skin during processing [16]. A study on the fiber misalignment commonly expected with traditional blade manufacturing techniques showed that random fiber misalignment in manufacturing shows small effects on the turbine blade even at excessive misalignment up to 10° [17–19]. However, significant differences were found for highly asymmetric and imbalanced laminates. Results showed that the glass-fiber-based skin was not as sensitive to misalignment as the carbon-fiber spars, particularly in the cap location (Figure 5.3).

5.2.1.5 Fiber Waviness (Out-of-Plane)

In general, process specifications do not permit ply waviness or wrinkles in a cured composite structure due to their significant effect on stiffness and strength properties. This section provides guidance on assessing the condition

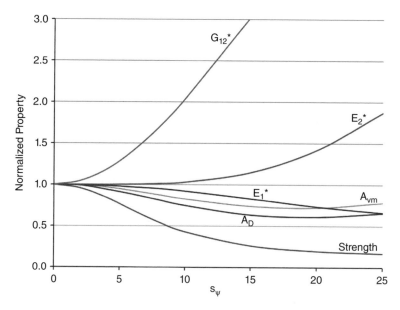

Figure 5.3 Effect of in-plane fiber misalignment (*x*-axis in degrees) on the graphite/epoxy composite strength, effective axial shear modulus G_{12}, transverse Young's modulus E_2, and axial Young's modulus E_1. A_{vm} and A_d are areas of damage initiation envelope areas based on Von Mises and Damage initiation models [15]. Source: Reproduced with permission from Elsevier.

of composite structure with wrinkles. The strength correction factors for laminate wrinkles must be characterized in terms of parameters such as length, width, depth, and laminate nominal analysis thickness. In addition, the percentage of plies wrinkled maybe used in some cases (Figure 5.4).

When an edge scan is available, these characterization parameters can be calculated by measuring points on edge scans. The points should be measured along the ply with the most severe defect in the laminate. Reasonable engineering judgment should be used in locating the measurement points on the ply exhibiting the most severe distortion. Most variability is typically associated with defining the length parameter. Other researchers have used the angle of the defect but that is not always possible. When an edge inspection is not possible, the other methods discussed previously can be used to obtain reasonable assessments of these characterization parameters. For example, the hyperspectral near infrared imaging can be used to obtain accurate assessments of resin pockets near the surface that can be associated with the "D" parameter of surface wrinkling. When all the parameters cannot be ascertained, a conservative approach shall be used and worse case scenarios need to be considered (Figure 5.5).

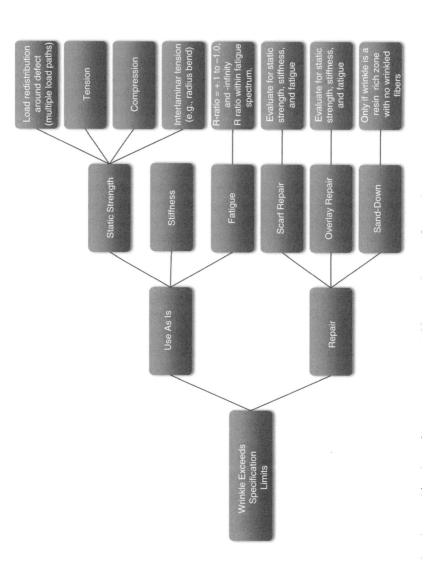

Figure 5.4 Engineering considerations when encountering wrinkles exceeding specification limits.

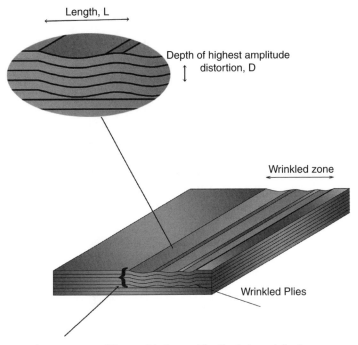

Length, L

Depth of highest amplitude distortion, D

Wrinkled zone

Wrinkled Plies

In some cases it is possible to consider % of plys wrinlked

Figure 5.5 Characterization of out-of-plane waviness in composites.

The measurements can be accomplished with the use of an image analysis program. The pixels for each point are measured and recorded. The defect should be characterized as fully as possible, including the depth D, and the length, L, of the worst wrinkled ply. The greatest magnitude defect should be used to characterize the defect location. It is also important to calculate the percentage of the laminate thickness that is wrinkled. This is calculated by determining the ratio of the total thickness of the plies that are wrinkled (typically $L/D < 15$ for unidirectional prepreg) to the laminate's design thickness. If one ply is wrinkled, it is common to proceed with the analysis without the affected ply. However, the mechanics of failure become significantly more complex as more layers are wrinkled. In certain analysis, the percentage of $0°$ plies relative to the total number of plies maybe also an important factor in determining the load carrying capacity. Also stacking sequence may be important, particularly the numbers of adjacent $0°$ plies having common wrinkles.

Wrinkles caused due to tooling at joints (e.g., at mandrel seams) or sharp changes in curvature can result in a portion of the laminate being wrinkled. Multiple wrinkles may also be present. The defect should be characterized as fully as possible, including the depth D, and the length, L, of the worst wrinkle in the series. It is also important to calculate the percentage of the laminate thickness that is wrinkled as discussed before. The greatest magnitude defect should be used to characterize the defect in the location of the composite structure. In the cases where a combination of fabric and unidirectional or tape plies are used, the reference surface shall be based on the tape plies. Sometimes wrinkles can occur in built-up structure using either cocured or partially cured members. These can occur at reinforcements occurring at noodle locations that are over- or underfilled. Typically, the surface ply in the structure to be reinforced is the most distorted by the noddle region. In composites using fabric materials, an additional challenge can occur since the material already contains an inherent amount of waviness. Post-cure resin indications of wrinkles can be an indicator of wrinkles when there is no cut edge to determine the wrinkle characteristics and can be coupled with the nondestructive testing methods described previously (Figures 5.6–5.8, Table 5.2).

5.2.1.6 Waviness in Curved Parts

Curved parts are sensitive to tooling, stacking sequence, part thickness, and resin properties. Resin migration, disbonds, fiber waviness, and porosity can all occur in the curved regions [19]. In the V-22 FSD (Full-Scale Development) program up to six composite fuselage sections were manufactured in which the curved frames were manufactured on male tools. Significant ply wrinkling

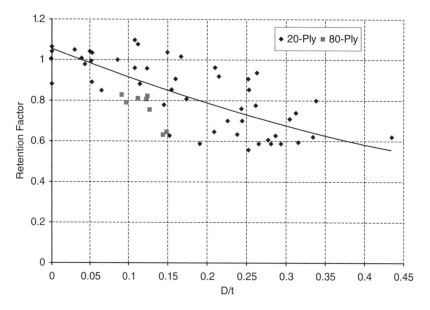

Figure 5.6 Compression strength retention factors versus wrinkle aspect ratio (D/t) for hybrid (uni + fabric) laminates.

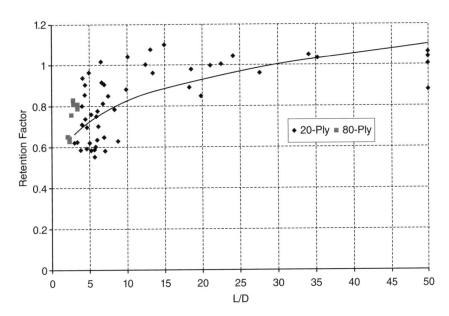

Figure 5.7 Compression strength retention factors versus wrinkle aspect ratio (L/D) for hybrid (uni + fabric) laminates.

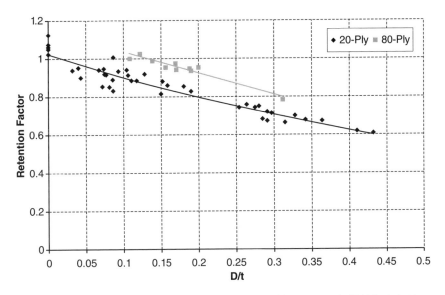

Figure 5.8 Tension strength retention factors versus wrinkle aspect ratio (D/t) for hybrid (uni + fabric) laminates.

Table 5.2 Analysis for fiber wrinkle defects.

Fiber misalignment or wrinkles	For 1 ply, strength and stiffness loss can be estimated by assuming loss of load carrying capacity due to the waviness. In compression, 10–25% reduction if surface ply is in 0° direction. For 1 ply, strength and stiffness loss can be estimated by assuming loss of load carrying capacity due to the waviness.	Refs. [17, 18].
	For more than 1 ply, Nonlinear dependence on D/L and % plies wrinkled. Higher impact on unidirectional layups.	
	In tension loading for more than 1 ply, Not as significant as matrix dominated compression loading but can initiate delamination at loads before ultimate that can then propagate under compression.	
	Fatigue can be an issue at certain strain levels, can reduce fatigue life by factor of 10.	

occurred in the frame radii after cure – probably as a result of the longitudinal and through-thickness thermal mismatch typical of composite laminates. Mabson and Neall [20] developed an analysis for predicting the interlaminar tension stresses in the frame radii due to a bending moment as applied in the testing. This paper reports on the numerous bending tests on the frame flanges to evaluate the structural effects of the wrinkles in the frame radii, and their structural acceptability. The results of frame flange bending tests were used to develop reductions in the interlaminar tension allowable due to the wrinkle depths in the frame radii (Figure 5.9).

5.2.1.7 Ply Gaps and Overlaps

These can normally occur in structures manufactured using hand or automated procedures. In tow or fiber placement method, they can be easier to characterize because of the repeated nature of the defect due to equipment-tool interaction. The effects of these on the structural properties of the V-22 EMD (engineering and manufacturing development) tow placed aft fuselage were previously reported in Chapter 2. Tension and compression failure strains were observed to decrease with increasing lap and or gap size. A reduction of 19% in the compression allowable strain was made to accommodate up to 25% of the aft fuselage skin laminates. It was commented that the strength reductions were probably caused by waviness in the 0° plies occurring as a result of the gaps and overlaps. Reductions of 6–17% in compression properties depending on layup with higher reductions for gaps in 90° plies has been reported. Some unnotched tension properties up to 13% maybe affected depending on degree [22–24]. Compression strength reductions between 5 and 27% were observed in laminates containing at least one 0.03-in. wide overlap or gap [25].

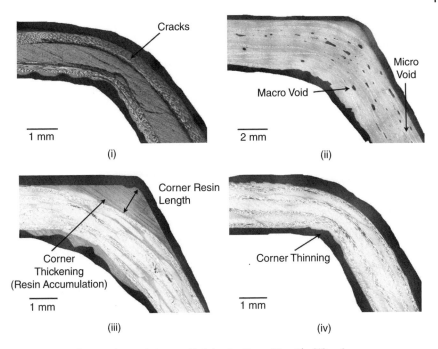

Figure 5.9 Defects in the angle beams (i) delaminations, (ii) voids, (iii) resin migration-corner thickening, and (iv) resin migration-corner thinning [21]. Source: Reprinted with permission from DEStech Publications, Inc.

The strength reductions did not increase significantly beyond these initial levels when wider 0.10-in. wide gaps were present or if several gaps were present. It is recommended that the effects of these imperfections be included in structural substantiation test articles if the process used results from these defects.

5.2.2 Cuts, Scratches, and Gouges

Custs, scratches, and gouges can reduce static strength by up to 50% [17, 26, 27]. Severed layers will reduce tension properties. An eccentric beam model can be used to estimate loss. Local delaminations are possible at the root of notches driven by loss of symmetry in the laminate. In addition, there is a possibility for moisture-driven delaminations.

Specimens with V-notch scratches of various lengths and depths were studied in graphite/epoxy composites [13]. The test specimens were 2 in. wide and the scratch lengths of 1.5 in. and 1 in. and were oriented normally to the applied uniaxial load. Scratches were found to be more detrimental in tension, particularly if they are deep or located along a specimen edge, and especially if the 0° plies are cut (Table 5.3). If the scratch is in the center of the specimen,

Table 5.3 Tension strength reductions for scratches in graphite/epoxy composite reported in the literature [28].

Defect (in 32-ply laminate)	% of basic laminate tensile strength
Center scratch: three plies deep	75.1
six plies deep	69.7
Edge scratch: three plies deep	75.1

approximately 40–50% reductions in tension strength were observed for the three- or six-ply scratches (in a 16-ply laminate). The strength reductions are more severe (50 vs. 40% reduction) for the edge versus center scratch. The compression results appear to be less severely affected than the tensile results due to the scratches. However, it should be noted that notched specimens were not considered in these studies. The behavior is not always intuitive. Open and filled holes specimens would have to incorporate damage at the hole perimeter. Damage in open-hole specimens loaded in tension can sometimes show an increase in strength, probably due to a relaxation of the notch effect, but in compression the reverse can be true.

5.2.3 Edge Delaminations

Delaminations are thought to be the most important cause for strength degradation in composites subject to impact. While it is not always possible to control service impacts, delaminations during manufacturing may grow, therefore it is important to carefully control of their size during manufacturing. In addition, composite parts may also be subject to edge delaminations during handling or moving.

Two-inch specimens were tested with simulated delaminations along one edge. Edge delaminations of 0.3 in. and 0.6 in. were simulated by embedding semicircular Teflon disks at the laminate quarter thicknesses [13]. The edge delaminations affected static compressive strengths more than static tensile strengths as expected with maximum reductions of 20% and 15%, respectively, for the larger delamination. Fatigue loading in compression caused some specimens with both delaminations to fail prematurely during the test.

5.2.4 Foreign Object Impact

Effects of impact damage can be seen in Table 5.4 on an older graphite/epoxy composite system, which nonetheless shows the effects of layup on residual

Table 5.4 Effect of foreign object impact on RTW strength (expressed as a percentage of unflawed strength) [13].

Laminate	Energy level	Tension strength (static), %	Tension residual strength (2 LT tension), %	Compression strength (static), %	Compression residual strength (2 LT compression), %
$(0/\pm45_2/0/\pm45)_s$	1	78	72	75	69
$(0_2/\pm45/0_2/90/0)_s$	1	71	77	60	55
$(0/\pm45/90)_{2s}$	1	91	51	70	35
$(0/\pm45/90)_{2s}$	2	42	30	61	31

Energy level 1 – incipient front face fiber failure; Energy Level 2 – incipient back face fiber failure.

Table 5.5 Effect of node disbonds on composite strength and failure modes (data from Hodge, A. and G. Dambaugh Ref. 70 Chapter 3).

Core batch	Density of disbonds detected	Failure mode	% of baseline capacity
1	None (Baseline)	Facesheet Delamination	100
2	Low	Core Shear	59
3	High	Core Shear	43

strength from different impact tests on composite laminates. In this study, two energy levels were used. In energy level 1, the amount of energy is modulated to cause incipient back face fiber failure while the impacted face, although indented, has no fiber failure (similar to barely visible impact damage (BVID)). Energy level 2 is at the threshold of causing front face fiber failure and substantial back face damage. Severe reductions are observed in strength before and after fatigue loading (two lifetimes in tension and compression spectra), both for the higher energy level and the level that was just barely visible on the nonimpacted face and virtually nondetectable (by visual examination) on the impacted face (energy level 1). Test results were also reported for graphite/epoxy sandwich beams and skin panel configurations subject to impact damage [28, 29]. The panels were exposed to random spectrum fatigue loads and then failed under static loading conditions. Testing conducted at room temperature with no moisture conditioning showed that BVID can reduce the strength to approximately 65% of the unflawed laminate strength for compressive loading and to approximately 85% for tensile loading. Note this is still above the notched properties used in design.

5.3 Effects on Sandwich Composites Properties

Sandwich design values are extremely sensitive to the following properties and damage:

- Core-facesheet bondline integrity
- Core splicing integrity
- Impact damage

Core-facesheet bondline integrity is an important factor in retaining both facesheet in-plane design strength, and sandwich panel flexural and shear strengths. The critical bondline failure modes are identified by Davis and Bond in reference [30] and shown in Figure 5.10. Additionally, they identified node bonds of the core as being critical to core integrity.

A study investigating the edgewise compression, flexure, and shear properties of small GFRP/PVC (glass fiber reinforced polymer/polyvinyl chloride) foam sandwich composite specimens found that determining the edgewise compression properties of a large sandwich structure using small specimens is difficult because the strength and failure mechanism are dependent on the gauge length [30, 31]. The compression strength decreases rapidly with increasing gauge length, and the failure mechanism changes from compressive fracture of the skins to shear crimping of the core.

5.3.1 Facesheet to Core Disbonding

When the sandwich structure is exposed to changes in temperature, humidity, and pressure, the airflow can be restricted, which results in large pressure differentials that can drive disbonds in the composite structure. This is especially the case for the spacecraft structures that are exposed to a combination of internal core pressure in the cells (relative to the external vacuum conditions existing in space), temperature, and possibly moisture-driven additional pressure if moisture is present within the core. Several honeycomb structural failures

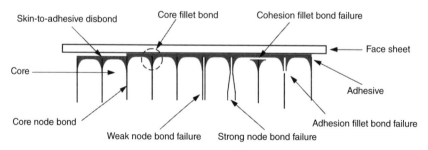

Figure 5.10 Core-facesheet adhesive bondline failure modes in sandwich structures. Source: From figure 6 in [30].

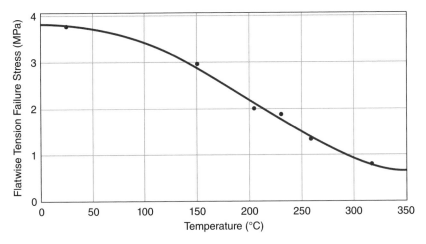

Figure 5.11 Effect of test temperature on facesheet to core bond strength, [35] Epstein and Ruth (1993).

have been reported associated with this failure mode in both civilian aircraft [32–34] and aerospace spacecraft systems [35]. As a result of these stresses, fully vented sandwich structures are typically used in space applications. Fracture mechanics approaches that consider approaches such as the Virtual Crack Closure Technique can be used to investigate the potential for a disbond to propagate in honeycomb structure [36]. Numerical simulations show that the disbond size is the most important parameter in evaluating whether a sandwich disbond will grow, followed by facesheet thickness and core thickness. Thinner facesheets were also found to cause more bulging that can increase the deformation ahead of the crack tip. Core thickness becomes an important parameter when thin facesheets are concerned. Proof testing can be considered for many cases, but test conditions need to account for the reduction in properties at elevated temperatures (Figure 5.11).

5.3.2 Facesheet Pillowing

Pillowing (see Chapter 3) is known to reduce the undamaged compression strength since it introduces a wrinkle in the facesheet. However, a study by Nettles [37] showed that while the knockdown in undamaged strength can be greater by 33% when pillowing is present, no such difference occurs when BVID is introduced. Precured to cocured sandwich laminates with four carbon-fiber/epoxy plies on an aluminum honeycomb core were investigated. When impacted with at 4.0 J mm^{-1}, the laminates showed the same compression strength. The study shows the importance of considering the interaction of defects and the importance of not considering one defect alone.

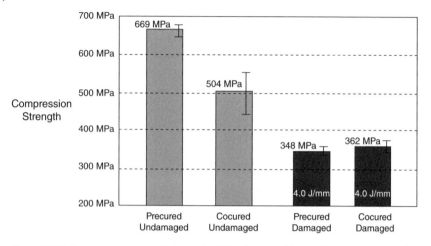

Figure 5.12 Compression strength in sandwich laminates without pillowing (pre-cured facesheets bonded to core) and with pillowing (cocuring of facesheets and core) [39]. (Image courtesy of NASA.)

The results show that in this case, the pillowing does not interact with impact; thus attempting to reduce the pillowing will increase compression strength when impact is considered (Figure 5.12).

5.3.3 Node Disbonds

In the case of aluminum cores used in sandwich construction, node disbonds can degrade the shear modulus of the core and can cause premature failure due to shear buckling, crimpling of the core or even by reducing the facesheet wrinkling resistance of the sandwich panel. Table 5.5 shows the effect of node disbonds from a study on carbon fiber epoxy composite sandwich structure with aluminum core showing varying degrees of disbonds in the core. It is expected that peeling nodes or misaligned nodes in flexible cell core geometries can also lead to reductions in these properties.

5.3.4 Core Splicing

Larger sandwich panels often require the core to be spliced. In the case of honeycomb cores, the splices involve nesting the cells as close as possible and then bonding with a foaming adhesive that expands and fills the cells' walls. If the core picking is not done properly, large gaps in between the cores can yield to a weaker bond between the two cores. The cores should also be inspected on both sides to ensure that there are no gaps. Table 5.7 shows some of the impacts associated with core splicing and other defects in sandwich structures.

Table 5.6 Impact of defects in sandwich composites.

Defect	Characterization feature	Key properties	Degradation
Core degradation due to water ingression	Material	Many	High degradation in core dominated properties. Can lead to skin-core disbond or corrosion if metal cores are used
Core splice: spacing exceeds limits, incomplete core splice	Density Change	Flexure, Compression (Edgewise) Thru' thickness shear	<10%
Cracked edge members	Structural Transition	Shear and Flexure	Critical structure for load transfer to sub-structure. High degradation possible
Crushed core at edge member	Structural Transition	Flexure, Compression (Edgewise)	Between 10 and 30%
Dents in facesheets	Mechanical	Flexure and Shear	Small dents can affect aero-properties. Larger dents can cause instabilities at high loads
Diagonal line of collapsed cells or nested cells	Density Change	Flexure, Shear (Flatwise), and Compression (Edgewise)	<10%
Drilled vent holes in skin	No-Density Change	Flexure	<10%
Edge close-out defects: gap between cores or core and edge member, voids in adhesive at edge, and incomplete edge seal	Structural Transition	Flexure, Compression (Edgewise), and Edge Bond	Can be >30%
Facesheet pillowing, wrinkling, or orange peel	Material	Flexure, Compression, and Shear	See discussion
Gaps in machined core/stepped skin	Structural Transition	Flexure, Compression	Between 10 and 30%

(continued)

Table 5.6 (Continued)

Defect	Characterization feature	Key properties	Degradation
Improper adhesive fillets	Material	Single Cantilever Beam	Instabilities and reduced fracture toughness
Incorrect core density	Material	Flexure, Shear (Flatwise), and Compression (Edgewise)	Poor shear strength or face instabilities
Porosity in facesheet	Material	Flexure, Compression (Edgewise)	27% with defect alone at 4–5% porosity but degradation goes up to 53% when considering impact [37]
			See discussion in laminates
Incorrect or variable core thickness	No-Density Change	Flexure, Shear (Flatwise)	Face instability or poor shear strength
Mismatched nodes or corrugations (in high density cores)	No-Density Change	Flexure, Shear (Flatwise), and Shear Fatigue	>30%
Moisture entrapment in facesheets	Material	Flexure, Shear (Flatwise)	Poor flexure strength
Over-expanded or blown core	No-Density Change	Flexure, Shear (Flatwise), and Shear Fatigue	Between 10 and 30%
Skin/core disbond	Material/ Mechanical	Flexure, Shear (Flatwise)	Instabilities and reduced fracture toughness
Misaligned ribbon or un-bonded nodes in core cell	Material (No-Density Change)/ Thermal	Flexure, Shear (Flatwise), and Shear Fatigue	Between 0% and 50%
Foreign object impact damage	Material	Compression	18% defect alone but degradation about 44% with impact [37]
			See discussion in laminates

Table 5.7 Effects of defects on joint strength [39].

	RTD tension	Compression 250°F
Out-of-round holes		
50/40/10 (by 0.004 in.)	–	–
30/60/10	−4.8%	–
Porosity around hole		
Severe	<2% effect	−30.8%
Tilted countersink		
10° Away from bearing surface	<2% effect	−18.7%
10° Toward bearing surface	−21.4%	−16.7%
Improper countersunk fastener seating depth[a]		
80% of thickness	−16.4%	–
100% of thickness	−34.3%	–
Interference fit		
50/40/10 @ 0.008 in.	<2% effect	+9.1%
30/60/10 @ 0.008 in.	<2% effect	<2% effect
Broken fibers on exit side of hole		
Delaminations extending 20–30% of laminate	−7.3%	−9.2%
Fastener removal and reinstallation [b]		
More than 100 cycles	<2% effect	−8.3%
Tilted countersink		
10° from bearing surface	Bearing strength, ∼−21% effect	Bearing strength, ∼−17%

a) Nominal was tested at 52% of laminate thickness. Simulate effects of dull bits and lack of backing material.
b) Fasteners were installed, torqued to 50 in.-pounds, and completely removed.

5.4 Effects on Bolted Joint Properties

Bolted joints are typically analyzed using a bearing-bypass methodology as discussed in Chapter 2. Figure 5.13 shows how coupon testing can be used to modify the bearing-bypass interaction. Defects should be addressed by examining their effects on the individual failure modes, that is, bypass tension and compression and bearing, followed by a test on the actual joint configuration. A study of seven manufacturing defects associated with mechanical fasteners

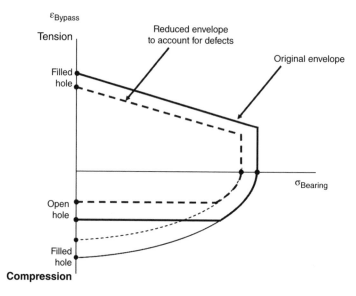

Figure 5.13 Modified bearing-bypass allowables in bolted joint analysis and design.

are reported in the literature [38]. These defects constitute out-of-round holes, broken fibers on the exit side of drilled holes, porosity, improper fastener seating depth, tilted countersinks, interference fit, and multiple fastener installation and removal cycles [39]. Test results of specimens from two laminates (50/40/10 and 30/60/10) indicated little sensitivity to out-of-round holes. The specimens were tested to failure at three environmental conditions: room temperature dry (RTD), room temperature wet (RTW), and elevated temperature wet (ETW) on Hercules AS/3501-6 carbon/epoxy test specimens. In fatigue testing, residual strengths were, in general, equal to or greater than non-fatigued specimen static strengths although some hole wear was observed. Hole wear maybe used as a criterion to limit the use of the joint. The porosity levels tested were typically not found to found to affect the fatigue peak bearing strength of the bolted joints.

Drilling operations can lead to defects. Drilling in carbon-fiber composites involves a large amount of tool wear, which results in variations of the cutting effectiveness of the drill. Drill point angle, thrust force, torque, speed of drilling, backing materials, and wear progression on the tool will all influence the damage zone around the hole. Several numerical models have been proposed to analyze this problem [40, 41].

5.4.1 Delaminations at the Holes

In drilling fastener holes, a common defect is a delamination at the exit side of the drilled hole; typically several plies into the laminate. This defect can

Figure 5.14 Delamination mechanisms: (a) peel-up delamination at entrance; (b) push-down delamination at exit [42], Durao et al. 2014. Source: Reprinted under the CC BY 3.0.

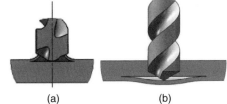

(a)   (b)

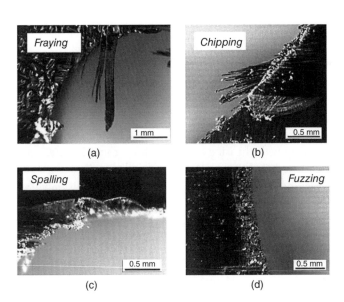

(a)   (b)

(c)   (d)

Figure 5.15 Typical defects associated with holes after drilling, (a) fraying; (b) chipping; (c) spalling; and (d) fuzzing Feito et al. 2014 [43]. Source: Reprinted under CC BY 3.0.

result because of improper drill speed, improper feed of the part to the drill, or improper backing of the composite part (Figure 5.14). The damage can also be considered in terms of peel-up and push-down damage induced loading modes. Peel-up is caused by the force as the drill is making contact with the specimen. The drill tends to abrade the materials that spiral upwards before being cut. This results in a peeling force that tends to separate the upper laminas of the plate. Similarly, push-down is a result of the compressive force as the drill is exiting the laminate. The uncut thickness becomes smaller and easier to bend, and thus induces delamination on the lower plies of the plate [42]. The result of drilling can also sometimes cause damage as shown in Figure 5.15. It is typical to assume a delamination zone near the hole due to these operations and free-edge stresses. Anecdotal evidence suggests that damage at the edge of holes in Kevlar and graphite laminates can result in higher tension strength. However, this may be a function of fiber treatments or resin properties.

A study of delamination near the holes for three configurations has been reported in the literature for (i) unreinforced holes, (ii) reinforced holes, and (iii) softened holes (0° graphite plies replaced with E-glass epoxy) [13]. The static results show that the delamination at a hole defect is more detrimental in compression than in tension. The reinforcement and softening methods appear to be advantageous for specific laminate configurations only. The residual strength results also show that the application of fatigue does not appear to materially increase the effect of the delamination as compared to the static results.

5.4.2 Oversize Holes

When holes are drilled too large an excessively loose fit may occur in the bolted joint (Figure 5.16). Static and fatigue testing conducted on specimens containing two-hole joints where one hole was sized properly and one hole had a fastener/hole clearance five times as large as the standard clearance was reported in the literature [13]. Changing the bearing load at the hole from 25 to 100% did not show an extreme sensitivity to the oversized hole. In all cases both the static and fatigue tests did not show extreme sensitivity to this manufacturing flaw.

5.4.3 Over-Torqued Fasteners

In the same study [13], the issue of fastener over-torque was also investigated. One set of fasteners in a 16 ply laminate were torqued to the 2.82 Nm (25 in-lb) standard and the other set was torqued to 7.90 Nm (70 in-lb). This value selected is based on the condition of the onset of noticeable fastener head dimpling, or to a value of torque at which pull-through of 0.127 mm (0.005 in) is experienced. Results of the tests show that unreinforced static tension

Figure 5.16 Example of an incorrectly drilled hole with improper tools in a carbon-epoxy composite laminate. Such defect can be analyzed as a larger hole.

Table 5.8 Tension strength reductions for defects near holes in graphite/epoxy composites [28].

Defect (in 32-ply laminate)	% of basic laminate tensile strength
Unloaded softened holes (softening using E-glass replacement fibers)	
No flaw	66.8
Delamination at hole	70.7
Over-torqued fastener	61.0
Loaded softened holes (during fatigue only)	
No flaw	72.3
Delamination at hole	62.5
Over-torqued fastener	71.4

strengths are degraded (approximately 10%) more than reinforced or softened static strengths because of the over-torqueing [13]. Fatigue exposure seemed to increase the sensitivity of the laminates to the presence of an over-torqued fastener (Table 5.8).

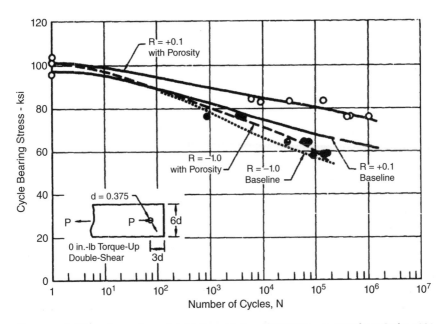

Figure 5.17 Effect of porosity on joint fatigue life in a 50/40/10 layup. Data from Garbo 1981 [44].

5.4.4 Porosity Near Fasteners

Tests of specimens with moderate porosity were conducted to evaluate the effects of this anomaly on joint durability [43, 44]. For the levels of porosity tested in regions near fastener holes showed a minor effect on static strength and joint fatigue life (Figure 5.17, Table 5.8).

5.5 Effects on Bonded Joint Properties

Test standards such as American Society for Testing and Materials (ASTM) D1002 [45], D3165 [46], and D5868 [47] can be used to study the effects of possible defects on the shear strength of bonded assemblies [48–51]. Other methods can also be used to evaluate oxygen degradation and fatigue [52, 53]. Those tests can also be useful as an in-process quality control approach for laminated assemblies by making the assembly oversized and cutting a portion of it for evaluation using this standard or fabricating a traveler panel from the same material batches. Caution should be used when trying to obtain design allowables from the coupon configurations because actual loading conditions may differ in the joint, and the process used to make the specimen may not be the same as that used to fabricate a large assembly in an autoclave. For more information on interpretation of the results, the reader is referred to ASTM D4896 [54]. The differences in the coefficient of thermal and hydroexpansion in the adhesive and adherends that further complicate the scaling efforts. However, the approach may be used to evaluate effects of deviations from the surface preparation procedures or environmental exposure to chemicals or thermal shock.

Defects in the adhesive joint near the free edges are typically the most detrimental. This zone tends to have a multiaxial stress state and it is also the region which is most exposed to environmental effects. The area near the free edges (at the location of load introduction) essentially supports the applied load and increasing the bond width beyond a certain area will have a negligible influence on the joint strength [55]. The details of the adhesive termination tend to be critical and must be addressed for the possibility of defects. The fillet should never be removed, since this has a beneficial effect in reducing stress concentrations at the joint ends. Controlling this fillet is critical for reducing joint stresses [56]. The adhesive thickness and stiffness also affects the ability of the spew fillet to reduce the stresses within the joints. A lower stiffness modulus relative to the adherends will tend to be more favorable. Thicker adhesive joints are likely to promote plane strain conditions, which typically are associated with more brittle behavior and tends to be less sensitive to the fillet configuration. Increase of adhesive thickness relative to adherend thickness tends to reduce adhesive shear stress, but may promote peel failures with increasing principal stresses.

Water is extremely detrimental to adhesives, particularly distilled water, which may cause more damage than salt water. Epoxides contain —OH groups that makes them polar, attracting water molecules that form hydrogen bonds with each other [57]. Water can cause degradation by plasticization, causing adhesives to crack, craze, or hydrolyze, thus attacking the interface or by causing swelling that causes additional stresses in the joint [57].

Bond durability is also dependent on the resistance of the adhesive-to-adherend interface degradation, which is a combination of environment and stress. ASTM D3762 [58] is a qualitative test that can be used in determining variations in adherend surface preparation parameters and adhesive environmental durability. It has been used to study durability of bonded joints, surface preparation effects and adhesives, and exposure under different environments [57, 59–63].

5.5.1 Assessment of Defects in Design of Bonded Joints

Defects that can be expected in the bonded such as porosity and wrinkles can be compensated for with the additional replacement plies applied to build up the necessary strength required. Designs that are based on the unnotched values will need to account for the reductions in strengths for those properties, whereas notched properties may be impacted less than the unnotched ones.

5.6 Statistical Considerations

In typical allowables development, as discussed in Chapter 2, we have shown how B-basis design allowables may be used in redundant structures where load redistribution can be expected to occur. Typically, in the aerospace industry, the procedures and methodologies for allowable mechanical properties are derived for metals using the Metallic Materials Properties Development and Standardization (MMPDS) handbook, or from *Composite Materials Handbook-17* (CMH-17) for the case of composites. While these approaches are widely accepted, these standards require large number of tests to be conducted and are not always feasible to implement when defects are concerned. These standards assume the process is controlled and is largely repeatable.

When a process disruption occurs that results in an imperfection or a defect, it is not always possible to perform an allowables test program based on those standards because of the high cost and time considerations involved. Another limitation of the standards approach is that material property development activities typically occur in early the development stages. For a listing of the relevant test standards that can be used in composite lamina or laminates, bolted joints, sandwich composites and bonded joints, please see Tables 5.9–5.12.

In some cases, the same kind of defect (e.g., wrinkles, porosity, etc.) may occur in many places throughout the structure. This will often require new

Table 5.9 Suggested tests that can be used for evaluating defects in composite lamina or laminates.

Organization	Standard Number	Title
ASTM	D2344	Standard Test Method for Short-Beam Strength of Polymer Matrix Composite Materials and Their Laminates
ASTM	D3039	Standard Test Method for Tensile Properties of Polymer Matrix Composite Materials
ASTM	D3171	Standard Test Methods for Constituent Content of Composite Materials
ASTM	D3410	Standard Test Method for Compressive Properties of Polymer Matrix Composite Materials with Unsupported Gage Section by Shear Loading
ASTM	D3479	Standard Test Method for Tension-Tension Fatigue of Polymer Matrix Composite Materials
ASTM	D3518	Standard Test Method for In-Plane Shear Response of Polymer Matrix Composite Materials by Tensile Test of a ±45° Laminate
ASTM	D3552	Standard Test Method for Tensile Properties of Fiber Reinforced Metal Matrix Composites
ASTM	D4255	Standard Test Method for In-Plane Shear Properties of Polymer Matrix Composite Materials by the Rail Shear Method
ASTM	D5229	Standard Test Method for Moisture Absorption Properties and Equilibrium Conditioning of Polymer Matrix Composite Materials
ASTM	D5379	Standard Test Method for Shear Properties of Composite Materials by the V-Notched Beam Method
ASTM	D5448	Standard Test Method for Inplane Shear Properties of Hoop Wound Polymer Matrix Composite Cylinders
ASTM	D5449	Standard Test Method for Transverse Compressive Properties of Hoop Wound Polymer Matrix Composite Cylinders
ASTM	D5450	Standard Test Method for Transverse Tensile Properties of Hoop Wound Polymer Matrix Composite Cylinders
ASTM	D5467	Standard Test Method for Compressive Properties of Unidirectional Polymer Matrix Composite Materials Using a Sandwich Beam
ASTM	D5687	Standard Guide for Preparation of Flat Composite Panels with Processing Guidelines for Specimen Preparation
ASTM	D6641	Standard Test Method for Compressive Properties of Polymer Matrix Composite Materials Using a Combined Loading Compression (CLC) Test Fixture
ASTM	D6856	Standard Guide for Testing Fabric-Reinforced "Textile" Composite Materials
ASTM	D7028	Standard Test Method for Glass Transition Temperature (DMA Tg) of Polymer Matrix Composites by Dynamic Mechanical Analysis (DMA)
ASTM	D7078	Standard Test Method for Shear Properties of Composite Materials by V-Notched Rail Shear Method
ASTM	D7264	Standard Test Method for Flexural Properties of Polymer Matrix Composite Materials

Table 5.10 Suggested tests that can be used for evaluating defects in structural joints.

Organi-zation	Standard Number	Title
ASTM	D5766	Standard Test Method for Open-Hole Tensile Strength of Polymer Matrix Composite Laminates
ASTM	D5961	Standard Test Method for Bearing Response of Polymer Matrix Composite Laminates
ASTM	D6264	Standard Test Method for Measuring the Damage Resistance of a Fiber-Reinforced Polymer-Matrix Composite to a Concentrated Quasi-Static Indentation Force
ASTM	D6484	Standard Test Method for Open-Hole Compressive Strength of Polymer Matrix Composite Laminates
ASTM	D6742	Standard Practice for Filled-Hole Tension and Compression Testing of Polymer Matrix Composite Laminates
ASTM	D6873	Standard Practice for Bearing Fatigue Response of Polymer Matrix Composite Laminates
ASTM	D7136	Standard Test Method for Measuring the Damage Resistance of a Fiber-Reinforced Polymer Matrix Composite to a Drop-Weight Impact Event
ASTM	D7137	Standard Test Method for Compressive Residual Strength Properties of Damaged Polymer Matrix Composite Plates
ASTM	D7248	Standard Test Method for Bearing/Bypass Interaction Response of Polymer Matrix Composite Laminates Using 2-Fastener Specimens
ASTM	D7332	Standard Test Method for Measuring the Fastener Pull-Through Resistance of a Fiber-Reinforced Polymer Matrix Composite
ASTM	D7615	Standard Practice for Open-Hole Fatigue Response of Polymer Matrix Composite Laminates

approaches to modify or redevelop the allowables/design values for the areas of structure where these occur. A suggested overall approach is shown in Figure 5.18. This approach may be used to develop a knockdown factor to be applied to legacy allowables. Generally, this factor is developed from tests, with typically less than 100 replicates taken over a range of laminate configurations. Hence the replicates at each configuration will be small and development of statistically significant data may be challenging. When such imperfections occur one approach that can be used is that from NASA, which includes the development of a material usage agreement (MUA) [64]. In this document, the hardware developer provides a plan that includes the technical rationale describing the material property development philosophy and provides detailed insight into how the material design properties are determined.

Table 5.11 Suggested tests for evaluating defects in sandwich composites.

Organi-zation	Standard Number	Title
ASTM	C271	Standard Test Method for Density of Sandwich Core Materials
ASTM	C272	Standard Test Method for Water Absorption of Core Materials for Sandwich Constructions
ASTM	C273	Standard Test Method for Shear Properties of Sandwich Core Materials
ASTM	C297	Standard Test Method for Flatwise Tensile Strength of Sandwich Constructions
ASTM	C363	Standard Test Method for Node Tensile Strength of Honeycomb Core Materials
ASTM	C364	Standard Test Method for Edgewise Compressive Strength of Sandwich Constructions
ASTM	C365	Standard Test Method for Flatwise Compressive Properties of Sandwich Cores
ASTM	C366	Standard Test Methods for Measurement of Thickness of Sandwich Cores
ASTM	C393	Standard Test Method for Core Shear Properties of Sandwich Constructions by Beam Flexure
ASTM	C394	Standard Test Method for Shear Fatigue of Sandwich Core Materials
ASTM	C480	Standard Test Method for Flexure Creep of Sandwich Constructions
ASTM	C481	Standard Test Method for Laboratory Aging of Sandwich Constructions
ASTM	D6416	Standard Test Method for Two-Dimensional Flexural Properties of Simply Supported Sandwich Composite Plates Subjected to a Distributed Load
ASTM	D6772	Standard Test Method for Dimensional Stability of Sandwich Core Materials
ASTM	D6790	Standard Test Method for Determining Poisson's Ratio of Honeycomb Cores
ASTM	D7249	Standard Test Method for Facing Properties of Sandwich Constructions by Long Beam Flexure
ASTM	D7250	Standard Practice for Determining Sandwich Beam Flexural and Shear Stiffness
ASTM	D7336	Standard Test Method for Static Energy Absorption Properties of Honeycomb Sandwich Core Materials
ASTM	D7766	Standard Practice for Damage Resistance Testing of Sandwich Constructions
ASTM	D7956	Standard Practice for Compressive Testing of Thin Damaged Laminates Using a Sandwich Long Beam Flexure Specimen
ASTM	D8067	Standard Test Method for In-Plane Shear Properties of Sandwich Panels Using a Picture Frame Fixture

Table 5.12 Suggested tests for evaluating defects in bonded joints.

Organi-zation	Standard Number	Title
ASTM	D1002	Standard Test Method for Apparent Shear Strength of Single-Lap-Joint Adhesively Bonded Metal Specimens by Tension Loading
ASTM	D 3762	Standard Test Method for Adhesive-Bonded Surface Durability of Aluminum (Wedge Test)

5.6.1 Mean versus Design Values

Typically the scatter in effects-of-defects test data is larger than those used in generating the original design allowables. The question arises as to whether "mean" values on retention factors should be used or whether new design allowables should be determined. In some cases, the retention curves using lower bound values may duplicate the scatter provided by basic design value strength allowables and maybe excessively conservative.

5.6.2 Simpson's Paradox

In the analysis of test results it is important to consider the Simpson's paradox, or sometimes called the Yule–Simpson effect or the amalgamation paradox [65]. The effect describes a condition that may occur when different groups may show one trend when analyzed separately, but when combining the data these trends may disappear or in fact reverse. Figure 5.19 shows an example where two sets of data, when partitioned, show one trend but when combined the trend reverses. The importance of understanding this principle is the importance of not drawing quick conclusions without understanding the true causes. Deciding on using aggregate or partitioned data requires careful consideration of the different variables and the ability to identify confounding variables. The best approach to guide against drawing a wrong conclusion is to have large data sets with representative laminates, stacking sequences, thicknesses, and defects across the various morphologies. Consequently, design of experiment approaches can be used to identify confounding variables and reduce the amount of testing required.

5.6.3 Design of Experiments

Screening tests using design of experiments (DOE) techniques can maximize the efficiency of testing and can be used to study sensitivity of the process to certain defects. The challenge with the screening tests is to decide the scale of the test, for example, coupon, element, and so on. Based on the results of

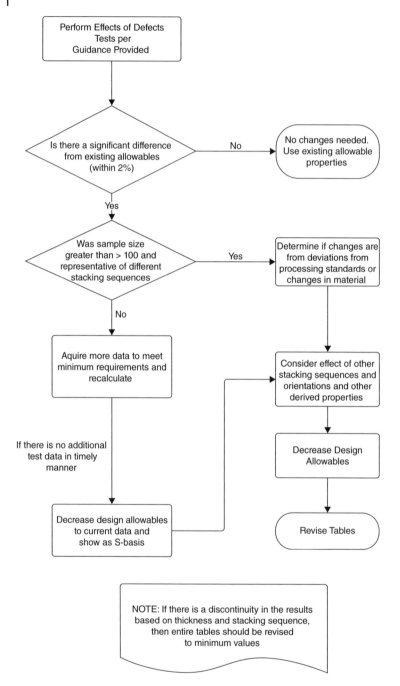

Figure 5.18 Suggested approach for evaluating the effects of defects on material allowables.

Figure 5.19 Simpson's paradox for quantitative data: a positive trend (—,—) appears for two separate groups, whereas a negative trend (----) appears when the groups are combined. Source: By Shultz 2007, public domain.

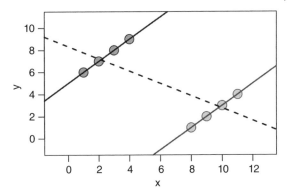

the screening tests, the range of parameters that can produce an acceptable product can be developed. DOE can be a powerful tool to determine the significance of a particular process or defect (input) and how it may influence certain mechanical properties (outputs). *A multifactor experiment* involves varying some of the controllable factors (e.g., layup, tooling type, design, material, etc.) in a prescribed manner to study their relationship with the output responses (e.g., mechanical property, defect morphology, etc.). The controllable factors are held constant while the uncontrollable factors (e.g., ambient conditions, variations in incoming materials, operators, etc.) are typically dealt with through blocking and randomization strategies. Typically, the controllable factors are treated as continuous, numerical variables, and are coded to factors between −1 and 1. Category factor levels can also be applied to the controllable factors. When conducting multifactor DOE, the assumption is that the true mean of an output variable is a function of the different controllable factors. The objectives of multifactor DOE is to study the relationship between the response and the factors or to identify the relationships that lead to the optimum predicted response. Even if the form of the function relating the different variables is not known, the experimental values used can be fitted with a low-order Taylor series (polynomial) approximation over the experimental region to meet the objectives required. Statistical DOE, coupled with engineering expertise, can provide an efficient strategy for controlling defects in the processing of composite materials. For more on DOE approaches and applications to composites, the reader is referred to the following Refs. [66–70].

5.7 Suggested Approach for Evaluation of Defects

An approach has been used in the MMPDS specifications [71] for incoming data that is significantly below the "design allowables" for metallic materials. The MMPDS handbook is the primary source for statistically based, design allowable properties for metallic materials and fastened joints. It is interesting that in the absence of sufficient data (i.e., sample size below 100); this specification suggests reducing the "design allowables" to the actual incoming data, and naming these "S-basis" (see Section 2.8 for the definition). Having the precedent set in the MMPDS specifications, this may be possible approach in the development of design data for composite structure having defects. However, care should be taken in the presence of extreme data variability, which may occur in composites effects-of-defects data, and this approach may be nonconservative. It is worth noting that metals data being referred to in the MMPDS normally exhibits lower data variability than composites data (see Chapter 1).

Typically, retention factors are determined from an effects-of-defects test program at the coupon level using the simulation of the actual defects and actual laminates. These can apply singularly to the manufacturing anomalies. Depending on the design approach, the retention factors for notched strength, stiffness, and fracture toughness properties are usually obtained. It is the responsibility of the structural analyst to ensure that they have captured the manufacturing anomaly effects on all the variables that may affect the results of their structural analysis. This task should also consider the possibility of multiple defects occurring at the same time. Considering the interaction of defects is critical. For example, porosity and impact damage may interact heavily whereas the interaction maybe much weaker if the impact damage occurs near a wrinkle. The suggested approach to evaluating defects is shown in Figure 5.18.

5.8   Evaluation of Scaling and Multiple Defects

Since defects are usually examined separately (e.g., through generation of test plans where a single defect is analyzed separately), the results do not account for interaction effects between defects. In the case of multiple defects degrading the resin properties (e.g., porosity, extensions of mechanical life, etc.), a cumulative retention factor, K_{tc} in the absence of test data can be expressed conservatively as:

$$K_{tc} = \prod_{i}^{n} K_{t,n} = K_{t,1} \cdot K_{t,2} \dots K_{t,n}$$

where n corresponds to number of defect types considered and $K_{t,n}$ is the retention factor for the nth defect. It may be an appropriate assumption that not all of the considered manufacturing effects are to be simultaneously occurring to the most severe degree. Therefore, the engineer should make an appropriate choice considering how to combine the various defects relating to their characterization and strength degradation and the likelihood to combine with the other defects in a given location. Propagation of damage from low energy impact is dependent on type of loading and strain levels. The strength loss of the damaged laminate can sometimes be approximated on the basis of an "equivalent" round hole [39]. However, this cannot be used in all cases when the damage is more extensive. Test results on stiffened panels indicate that impact damage produced an effective strain concentration greater in magnitude than a round hole of equivalent size and this approach can be nonconservative. Strengths predicted for an equivalent hole size were not conservative for non-stiffened panels by approximately 30%, possibly due to local structural instability of delaminated plies within the damaged zone; however, in stiffened panels this reduction was not observed [39].

Limited studies have directly addressed the scaling effects on all possible composite defects. One study reported a reduction in apparent strength up to 20% in going from laboratory specimens to full-scale structure for carbon/epoxy rocket cases [72]. This strength reduction may be attributed to a possible scale effect associated with the increase of stressed material volume or in changes in test procedures or manufacturing methods. It is believed that manufacturing imperfections associated with making larger structures can account for the impact on some of the properties observed. For example, Verette and Labor [73] studied the scale effects on mechanical properties in fiber-dominated graphite/epoxy solid laminates under static and spectrum fatigue uniaxial loading. The large size specimens had a test volume 16 times that of the small size specimens. The scale effects reduction of large-scale versus small-scale specimens was found to be −4.5% for static strength, −8.0% for tension dominated spectrum fatigue residual strength, and −3.2% for compression dominated spectrum fatigue residual strength. For bonded joints tested under RTW with fiber-dominated adherends, no significant scale effect strength reductions were observed for tension and compression static strength and also for compression dominated fatigue residual strength [74]. However, in RTW tension dominated fatigue bonded joints, a significant lifetime reduction scale effect is observed. No significant scale effects strength reductions are observed for static tension and tension dominated fatigue residual strength of RTW bolted joints.

The scale effects are believed to occur because of quality issues associated with making larger structures compared to test coupon size specimens. If proper materials, consistent processing, and tooling issues are controlled, it is not expected that scale effects would be an issue. Scale effects may be observed due to design challenges for example when large thermal stresses develop when multiple materials are involved or in a multiple load path structure that was not validated with appropriate building block type testing. In the cases where size effects may be a factor, the equation proposed in [73] may be considered,

$$\frac{\beta_l}{\beta_s} = \frac{1}{(V_L/V_s)^{1/\alpha}}$$

where β is the characteristic strength for the large or small specimen, V is the volume of the large or small specimen, and α is the Weibull Shape parameter. An experimental program has been carried out on plates of carbon/epoxy laminates in different sizes [75]. In a series of experiments and analysis, the plates and impactors were scaled by a factor of five in size. The results show that the delamination determined by C-scan appeared to be more extensive for the larger specimens but consistent with fracture mechanics approaches for size effects. The results showed a similar delamination area from impact in different size plates, provided the impact velocity varies inversely with the square root of plate length.

If fatigue spectra were used a conservative approach would have to be scaling the maximum strain to that at limit load. In commercial aerospace structures, this strain level could be even lower, typically 70% of limit load. Generally, the static compressive strength of bolted structures drops with increasing damage but tension strengths may actually increase in some cases. Bearing damage should be monitored in fatigue loading, where hole elongation should be monitored (see ASTM standard D6873). The undamaged case tends to yield the most conservative case when examining the fatigue life of compression dominated structures where several studies have shown increased residual strength after fatigue of impacted specimens [76].

References

1 Marrouze, J., J. Housner, and F. Abdi, Effect of Manufacturing Defects and their Uncertainties on Strength and Stability of Stiffened Panels . ICCM19, Montreal, Canada, July 28–August 2, 2013.

2 Qamhia, I., E. Lauer-Hunt, and R. Elhajjar, Identification of acoustic emissions from porosity and waviness defects in continuous fiber reinforced composites. Advances in Civil Engineering Materials, 2013. 2(1): pp. 37–50.

3 Yang, P. and R. El-Hajjar, Porosity defect morphology effects in carbon fiber-epoxy composites. Polymer-Plastics Technology and Engineering, 2012. 15(11): pp. 1141–1148.

4 Zhang, A., D. Li, H. Lu, and D. Zhang, Qualitative separation of the effect of voids on the bending fatigue performance of hygrothermal conditioned carbon/epoxy composites. Materials and Design, 2011. 10: pp. 4803–4809.

5 Kastner, J., B. Plank, D. Salaberger, and J. Sekelja, Defect and porosity determination of fibre reinforced polymers by X-ray computed tomography. In 2nd International Symposium on NDT in Aerospace 2010. 2010.

6 De Andrade Silva, F., J.J. Williams, B.R. Muller, M.P. Hentschel, P.D. Portella, and N. Chawla, Three-dimensional microstructure visualization of porosity and Fe-rich inclusions in SiC particle-reinforced Al ally

matrix composites by X-ray synchrotron tomography. Metallurgical and Materials Transaction A, 2010. 41(8): pp. 2121–2128.

7 Hardin, R.A. and C. Beckermann, Effect of porosity on the stiffness of cast steel. Metallurgical and Materials Transactions A, 2007. 38(12): pp. 2992–3006.

8 Lopes, C.S., J.C. Remmers, and Z. Gurdal, Influence of Porosity on the Interlaminar Shear Strength of Fibre-Metal Laminates. American Institute of Aeronautics and Astronautics, 2006.

9 Hagstrand, P.O., F. Bonjour, and J.A.E. Mânson, The influence of void content on the structural flexural performance of unidirectional glass fibre reinforced polypropylene composites. Composites Part A: Applied Science and Manufacturing, 2005 . 36(5): pp. 705–714.

10 Costa, M.L., S.F.M. De Almeida, and M.C. Rezende, The influence of porosity on the interlaminar shear strength of carbon/epoxy and carbon/bismaleimide fabric laminates. Composites Science and Technology, 2001. 61(14): pp. 2101–2108.

11 Daniel, I.M., S.C. Wooh, and I. Komsky, Quantitative porosity characterization of composite materials by means of ultrasonic attenuation measurements. Journal of Nondestructive Evaluation, 1992. 11(1): pp. 1–8.

12 Hsu, D.K. and K.M. Uhl, A morphological study of porosity defects in graphite-epoxy composites, in Review of Progress in Quantitative Nondestructive Evaluation. 1987, Springer. pp. 1175–1184.

13 Verette, R.M. and E. Demuts, (1976). Effects of manufacturing and in-service defects on composite materials. in Proc. Army Symposium on Solid Mechanics. DTIC Document.

14 Wimmer, G., W. Kitzmüller, G. Pinter, et al., Computational and experimental investigation of delamination in L-shaped laminated composite components. Engineering Fracture Mechanics, 2009. 76(18): pp. 2810–2820.

15 Bednarcyk, B.A., J. Aboudi, and S.M. Arnold, The effect of general statistical fiber misalignment on predicted damage initiation in composites. Composites Part B: Engineering, 2014. 66: pp. 97–108.

16 Vanskike, W.P. and R.D. Hale. Effects of fiber misalignment on composite wind turbine rotor blades. in 35th Wind Energy Symposium. 2017.

17 Noor, A.K., M. Shuart, J. Starnes Jr, et al. Failure analysis and mechanisms of failure of fibrous composite structures. in NASA Conference Publication 2278. 1982. Hampton, VA: Nasa Langley Research Center.

18 Elhajjar, R.F. and S.S. Shams, A new method for limit point determination in composite materials containing defects using image correlation. Composites Science and Technology, 2016. 122: pp. 140–148.

19 Ma, Y., T. Centea, and S.R. Nutt, Defect reduction strategies for the manufacture of contoured laminates using vacuum bag-only prepregs. Polymer Composites, 2015. 38: pp. 2016–2025

20 Mabson, G. and P. Neall III, Analysis and Testing of Composite Aircraft Frames for Interlaminar Tension Failure. Rotary Wing Test Technology, 1988: pp. 1988.

21 Ma, Y., T. Centea, G. Nilakantan, et al., Vacuum Bag Only Processing of Complex Shapes: Effect of Corner Angle, Material Properties and Processing Conditions. 2014.

22 Altman, J., *Advanced Composite Serviceability Program*. AF Contract No. F33615–76-C-5344 (Rockwell International Corp./IITRI/Washington U.), Quarterly Progress Reports 1–9 (Jan. 1977 to Dec. 1978), 1980.

23 Ryder, J. and E. Walker, The effect of compressive loading on the fatigue lifetime of graphite/epoxy laminates. 1979, DTIC Document.

24 Croft, K., L. Lessard, D. Pasini, et al., Experimental study of the effect of automated fiber placement induced defects on performance of composite laminates. Composites Part A: Applied Science and Manufacturing, 2011. 42(5): pp. 484–491.

25 Sawicki, A. and J. Minguet. The effect of intraply overlaps and gaps upon the compression strength of composite laminates. in Collection of Technical Papers— AIAA/ASME/ASCE/SHS/ASC Structures, Structural Dynamics & Materials Conference. 1998.

26 Petersen, D.R., R.F. El-Hajjar, and B.A. Kabor, On the tension strength of carbon/epoxy composites in the presence of deep scratches. Engineering Fracture Mechanics, 2012. 90: pp. 30–40.

27 Shams, S.S. and R.F. El-Hajjar, Overlay patch repair of scratch damage in carbon fiber/epoxy laminated composites. Composites Part A: Applied Science and Manufacturing, 2013. 49: pp. 148–156.

28 Labor, J. and R. Verette, Environmentally controlled fatigue tests of composite box beams with built-in flaws. Journal of Aircraft, 1978. 15(5): pp. 257–263.

29 Labor, J., Impact damage effects on the strength of advanced composites, in Nondestructive Evaluation and Flaw Criticality for Composite Materials. 1979, ASTM International.

30 Davis, M.J. and D.A. Bond, The importance of failure mode identification in adhesive bonded aircraft structures and repairs. ICCM 12, Paris

31 Mouritz, A. and R. Thomson, Compression, flexure and shear properties of a sandwich composite containing defects. Composite Structures, 1999. 44(4): pp. 263–278.

32 Canada, T.S.B.o., Aviation Investigation Report, Loss of Rudder in Flight Air Transat, Airbus A310–308 C-GPAT Miami, Florida. 2007, The Transportation Safety Board of Canada, Quebec, Canada.

33 Air Accident Investigation Branch. AAIB Bulletin 8/92 Ref: EW/A92/5/1, A.A.I., UK, 1992.

34 Air Accident Investigation Branch. AAIB Bulletin 2/95 Ref: EW/C94/8/3, A.A.I., UK, 1995.

35 Epstein, G. and S. Ruth, Honeycomb sandwich structures: Vented versus unvented designs for space systems. 1993, DTIC Document.

36 Rinker, M., R. Krueger, and J. Ratcliffe, Analysis of an Aircraft Honeycomb Sandwich Panel with Circular Face Sheet/Core Disbond Subjected to Ground-Air Pressurization. 2013.

37 Nettles, A.T., Some Examples of the Relations Between Processing and Damage Tolerance. NASA 2–2131 2012.

38 Garrett, R.A., Effect of manufacturing defects and service-induced damage on the strength of aircraft composite structures. in Composite Materials: Testing and Design (Seventh Conference). 1986. ASTM International.

39 Garrett, R., Effect of defects on aircraft composite structures. 1983, DTIC Document.

40 Feito, N., J. López-Puente, C. Santiuste, et al., Numerical prediction of delamination in CFRP drilling. Composite Structures, 2014. 108: pp. 677–683.

41 Phadnis, V.A., F. Makhdum, A. Roy, et al., Drilling in carbon/epoxy composites: experimental investigations and finite element implementation. Composites Part A: Applied Science and Manufacturing, 2013. 47: pp. 41–51.

42 Durão, L.M.P., J.M.R. Tavares, V.H.C. De Albuquerque, et al., Drilling damage in composite material. Materials, 2014. 7(5): pp. 3802–3819.

43 Feito, N., J. Díaz-Álvarez, A. Díaz-Álvarez, et al., Experimental analysis of the influence of drill point angle and wear on the drilling of woven CFRPs. Materials, 2014. 7(6): pp. 4258–4271.

44 Garbo, S.P. and J. Ogonowski, Effect of Variances and Manufacturing Tolerances on the Design Strength and Life of Mechanically Fastened Composite Joints. Volume 1. Methodology Development and Data Evaluation. 1981, DTIC Document.

45 ASTM, D1002 Standard Test Method for Apparent Shear Strength of Single-Lap-Joint Adhesively Bonded Metal Specimens by Tension Loading (Metal-To-Metal). 2010. ASTM International: West Conshohocken, PA.

46 ASTM, D3165–07 Standard Test Method for Strength Properties of Adhesives in Shear by Tension Loading of Single-Lap-Joint Laminated Assemblies. 2007. ASTM International: West Conshohocken, PA.

47 ASTM, D5868 Standard Test Method for Lap Shear Adhesion for Fiber-Reinforced Plastic (FRP) Bonding. ASTM International: West Conshohocken, PA. 2001.

48 Vijayakumar, R., M. Bhat, C. Murthy, et al., Non destructive evaluation of adhesively bonded carbon fiber reinforced composite lap joints with varied bond quality. in AIP Conference Proceedings-American Institute of Physics. 2012.

49 Pantelakis, S., G. Gkikas, A. Paipetis, et al., Corrosion and environmental degradation of bonded composite repair. International Journal of Structural Integrity, 2013. 4(1): pp. 67–77.

50 Stratford, T. and J. Chen, Designing for tapers and defects in FRP-strengthened metallic structures. in Proceedings of the International Symposium on Bond Behaviour of FRP in Structures (BBFS 2005) Chen and Teng (Eds). 2005.

51 Kessler, M.R., R.H. Walker, D. Kadakia, et al., Evaluation of carbon/epoxy composites for structural pipeline repair. in 2004 International Pipeline Conference. 2004. American Society of Mechanical Engineers.

52 ASTM, D3632 Standard Test Method for Accelerated Aging of Adhesive Joints by the Oxygen-Pressure Method. 2011. ASTM International: West Conshohocken, PA.

53 ASTM, D3166 Standard Test Method for Fatigue Properties of Adhesives in Shear by Tension Loading (Metal/Metal). 2012. ASTM International: West Conshohocken, PA.

54 ASTM, D4896 Standard Guide for Use of Adhesive-Bonded Single Lap-Joint Specimen Test Results. 2016. ASTM International: West Conshohocken, PA.

55 Taib, A.A., R. Boukhili, S. Achiou, et al., Bonded joints with composite adherends. Part I. Effect of specimen configuration, adhesive thickness, spew fillet and adherend stiffness on fracture. International Journal of Adhesion and Adhesives, 2006. 26(4): pp. 226–236.

56 Tsai, M. and J. Morton, The effect of a spew fillet on adhesive stress distributions in laminated composite single-lap joints. Composite Structures, 1995. 32(1): pp. 123–131.

57 Adams, R., J. Cowap, G. Farquharson, et al., The relative merits of the Boeing wedge test and the double cantilever beam test for assessing the durability of adhesively bonded joints, with particular reference to the use of fracture mechanics. International Journal of Adhesion and Adhesives, 2009. 29(6): pp. 609–620.

58 ASTM, D3762 Standard Test Method for Adhesive-Bonded Surface Durability of Aluminum (Wedge Test). 2010. ASTM International: West Conshohocken, PA.

59 Armstrong, K., Effect of absorbed water in CFRP composites on adhesive bonding. International Journal of Adhesion and Adhesives, 1996. 16(1): pp. 21–28.

60 Armstrong, K., Long-term durability in water of aluminium alloy adhesive joints bonded with epoxy adhesives. International Journal of Adhesion and Adhesives, 1997. 17(2): pp. 89–105.

61 Thiedmanu, W., F. Tolan, P. Pearce, et al., Silane coupling agents as adhesion promoters for aerospace structural film adhesives. The Journal of Adhesion, 1987. 22(3): pp. 197–210.

62 Underhill, P. and D. DuQuesnay, The dependence of the fatigue life of adhesive joints on surface preparation. International Journal of Adhesion and Adhesives, 2006. 26(1): pp. 62–66.

63 Speth, D.R., Y.P. Yang, and G.W. Ritter, Qualification of adhesives for marine composite-to-steel applications. International Journal of Adhesion and Adhesives, 2010. 30(2): pp. 55–62.

64 NASA, NASA-STD-6016 Standard Materials and Processes Requirements for Spacecraft.

65 Blyth, C.R., On Simpson's paradox and the sure-thing principle. Journal of the American Statistical Association, 1972. 67 (338): pp. 364–366.

66 Montgomery, D.C. and D.C. Montgomery, Design and Analysis of Experiments. Vol. 7. 1984: New York, John Wiley & Sons, Inc.

67 Lipson, C. and N.J. Sheth, Statistical Design and Analysis of Engineering Experiments (Book-Statistical Design and Analysis of Engineering Experiments.). 1973: New York, McGraw-Hill Book Co.

68 Mohan, N., A. Ramachandra, and S. Kulkarni, Influence of process parameters on cutting force and torque during drilling of glass–fiber polyester reinforced composites. Composite Structures, 2005. 71(3): pp. 407–413.

69 Li, Z., Y. Jiao, T. Deines, et al., Rotary ultrasonic machining of ceramic matrix composites: feasibility study and designed experiments. International Journal of Machine Tools and Manufacture, 2005. 45(12): pp. 1402–1411.

70 Palanikumar, K., L. Karunamoorthy, and R. Karthikeyan, Assessment of factors influencing surface roughness on the machining of glass fiber-reinforced polymer composites. Materials and Design, 2006. 27(10): pp. 862–871.

71 MMPDS, Metallic Materials Properties Development and Standardization. 2012.

72 Swanson, S.R., Strength design criteria for carbon/epoxy pressure vessels. Journal of Spacecraft and Rockets, 1990. 27(5): pp. 522–526.

73 Verette, R. and J. Labor, Structural criteria for advanced composites. AFFDL-TR-76-142, Vol. I, Final Report,, 1977(F33615–74).

74 Jeans, L., G. Grimes, and H. Kan, Fatigue Spectrum Sensitivity Study for Advanced Composite Materials. US Air Force Flight Dynamic Laboratory, Technical Report AFWAL-TR-80-3130, Volumes I, II and III, 1980.

75 Qian, Y., S. Swanson, R. Nuismer, et al., An experimental study of scaling rules for impact damage in fiber composites. Journal of Composite Materials, 1990. 24(5): pp. 559–570.

76 Nettles, A.T., A. Hodge, and J. Jackson, Simplification of Fatigue Test Requirements for Damage Tolerance of Composite Interstage Launch Vehicle Hardware. 2010.

6

Selected Case Studies in Effects of Defects

6.1   Introduction

In this chapter, we present several case studies associated with failures in manufacturing or repair of composite structures that have been connected to defects in materials, manufacturing, or design. We have also attempted to go beyond highlighting the defect, but to also discuss any lessons learned that can be extended to other situations.

6.2   Case Study 1: The Ohio Timber Road II Wind Turbine Failure Due to Wrinkles

6.2.1   Event

On April 12, 2012 at approximately 12:48 p.m., Eastern Daylight Time, two 49 m blades on one Vestas v100 – 1.8 MW wind turbine broke at Timber Road II near Payne, Ohio in the United States [1]. Based on eyewitness accounts and data analysis, the incident was initiated when a single blade broke and struck the tower when rotating. The first incident of the rotor blade failure was then followed by another blade failure that scattered debris down to the surrounding area.

6.2.2   Background

The rotor blade is typically made from a composite spar (a reinforcing member) bonded to provide support to a composite skin. In the wind energy sector, the industry follows internal guidelines or those by some certifiers like DNV-GL [2]. This DNV-GL standard provides principles and technical requirements for rotor blades for wind turbines onshore and offshore.

Composite Structures: Effects of Defects, First Edition.
Rani Elhajjar, Peter Grant and Cindy Ashforth.
© 2019 John Wiley & Sons Ltd. Published 2019 by John Wiley & Sons Ltd.

6.2.3 Investigation

As part of the investigation, the manufacturer performed a multitude of coupon and structure level tests for tension strength and modulus, thickness and ply orientation, void and interface assessments, glass winding, resin cure state, and glass transition temperatures. Vestas concluded in their investigation that the root cause of the failure of the initial blade was a wrinkle or fiber misalignment in the carbon fibers of the spar [1]. The wrinkle caused damage to propagate to the point of failure after the blade experienced high loads for a low number of cycles. The blade was within its normal life and it is believed that the large wrinkle was an inspection escape. The second blade failure was determined to be due to the overload caused by the failure of the first blade. In summary, the investigation revealed that low cycle fatigue caused the failure. The failure was initiated at the wrinkle and progressed by a few cycles of high load amplitude (Figure 6.1).

A full blade test was conducted on another blade with the same defect and it showed that the blade was not capable of sustaining high loads after a few number of cycles. Using the acoustic emission test method, it was possible to determine the load level that corresponds to the initiation and then growth of damage. Operating the turbine below the damage initiation load, it was possible to show that damage does not grow.

6.2.4 Lessons Learned

A wrinkle defect present in a composite can make the composite structure sensitive to low cycle fatigue. The case highlighted the challenges in adequately screening for wrinkle defects during manufacturing. A laser shearography method was developed to capture wrinkle defects on other blades to detect defects with aspect ratios (L/D) of less than 10, a size they considered critical for damage growth. A precured patch was bonded to blades where these defects were found.

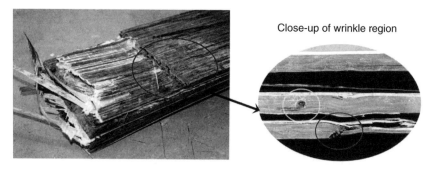

Close-up of wrinkle region

Figure 6.1 Tip section of wind turbine blade and close-up of wrinkled region [1].

6.3   Case Study 2: Faulty Repairs of Sandwich Core Structure

This case study appeared in a Federal Aviation Administration (FAA) sponsored study in cooperation with a major airline to document case studies of faulty field repairs associated with composite components [3].

6.3.1   Event

In this case, a commercial transport inboard flap was observed to have significant delamination between the metal skins and honeycomb core. A repair was performed incorrectly before proper approvals were made and the repair station had to remove the skin for an assessment. During the removal of the skin many processing related defects were observed. The following images show some of the processing defects. For more complete details about this study, the reader is referred to the original reference (Figures 6.2–6.4).

6.3.2   Lessons Learned

The many processing associated defects with the repair meant that the inboard flap assembly had to be reworked. In this case, there was no evidence of proper skin removal, proper core placement, or maintaining the bonding tolerance of

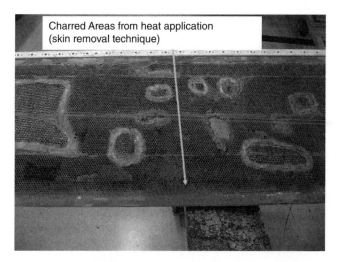

Charred Areas from heat application
(skin removal technique)

Figure 6.2 Damaged honeycomb core from excessive heating during removal of original skin [3].

Figure 6.3 Uneven film adhesive thicknesses may have been caused by tooling contour inaccuracies. Areas at the inboard and outboard ends had evidence of film adhesive porosity (small voids), which may indicate the film adhesive had a thickness that exceeded tolerances. This suggested that the tool used in the repair did not match the flap assembly correctly [3].

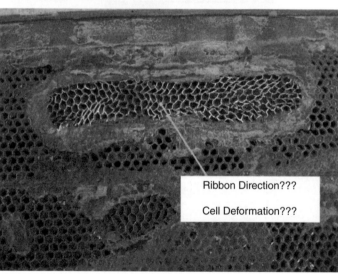

Figure 6.4 Distorted honeycomb core during installation. The technician may have squeezed the replacement core into the cut out area with no regard for orientation, shape, and cell wall condition [3].

the adhesive. In the end, the flap was rebuilt by the airline by replacing all the honeycomb core and skins.

6.4   Case Study 3: Bonded Repair Failure

This case study appeared in a FAA sponsored study in cooperation with a major airline to document case studies of faulty field repairs associated with composite components [3].

6.4.1   Event

In this case, the disbonding defect was discovered when a passenger looked out a window of the aircraft and observed severe damage to an outboard flap. Upon landing, it was discovered that approximately 80% of the trailing edge wedge assembly was missing from skins that had disbonded from the spar (Figure 6.5).

6.4.2   Investigation

Prior to this inflight incident, the airline sent the outboard flap assembly for repair. The investigation revealed the repair procedure for surface preparation of the bonding surface was not approved by the original equipment manufacturer. An alternative surface preparation procedure, chromate conversion coating, was used without specific work instructions instead of the approved phosphoric acid anodizing treatment.

6.4.3   Lessons Learned

Any repair procedure must be assessed and adequately characterized before application and this in the case of aircraft requires approval by both the original equipment manufacturer and any authorized oversight engineering representative. Any repair must also have the goal of producing a good bond that can be demonstrated by proof testing or nondestructive testing. The wedge testing in American Society for Testing and Materials (ASTM) D 3762 Standard Test Method for Adhesive-Bonded Surface Durability also known as the Wedge Test can be used to increase the build reliability by bonding witness specimens in-parallel with the bonded repairs. In this case, the disbonding defect was discovered when a passenger looked out a window of the aircraft and observed severe damage to an outboard flap.

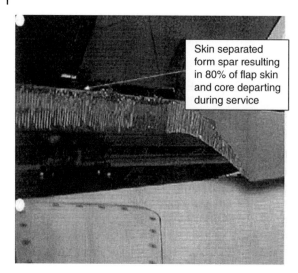

Skin separated form spar resulting in 80% of flap skin and core departing during service

Figure 6.5 Disbonding between skin and spar following improper repair [3].

6.5 Case Study 4: Air Transat 961 Sandwich-Composite Failure

6.5.1 Event

On March 6, 2005 an Air Transat Flight 961 was flying from Varadero, Cuba to Quebec City, Canada. When the aircraft was about 90 nautical miles south of Miami, Florida, United States, and in level flight, the flight crew heard a loud bang and felt some vibrations. The aircraft then experienced a Dutch roll, which the pilots controlled and then landed the aircraft back safely in Cuba. The aircraft experienced a significant structural failure which resulted in large portions of the rudder detached in flight (Figure 6.6).

6.5.2 Investigation

The investigation into the incident found that a disbond or in-plane core fracture existed in the rudder before the flight. The damage could have been caused by either a discrete service event or due to a weak bond at the z-section that is used to trim the front and bottom edge of each rudder. When curing the sandwich structure, dimensional tolerances at the end details (including the z-section) can result in low bonding pressure. Specimens extracted from near the z-section showed a meniscus consistent with insufficient bonding pressure. The damage to the rudder is thought to have grown over several cycles until the occurrence flight where it reached a critical size and then resulted in the loud and sudden explosion of the skin [4].

Figure 6.6 Right-side view of vertical tail plane and rudder residuals. Source: Courtesy of the Transportation Safety Board of Canada .

6.5.3   Lessons Learned

The investigation determined that the manufacturer's recommended inspection program was not adequate to detect all the rudder defects and that the damage may have been present for many flights before the incident. Selecting the correct nondestructive technique and its application frequency are essential for catching critical defects. The model of the rudder was also found to not include any design features that can mechanically arrest the growth of disbond damage or in-plane core failure before it reaches the critical size [4]. The structure should have redundancies or design features that would arrest a disbond such that it does not propagate in an unstable or unpredictable manner.

6.6 Case Study 5: Debonding Failure of a Sandwich-Composite Cryogenic Fuel Tank

6.6.1 Event

Sandwich based composite structures were developed for liquid hydrogen (LH2) tanks on the X-33 vehicle. The X-33 vehicle was a program by Lockheed Martin corporation for an unmanned, sub-scale technology demonstrator for suborbital spaceplane developed in the 1990s. The program was setup under the US government-funded Space Launch Initiative program. During prototype testing, the structure experienced an unexpected failure at the NASA Marshall Space Flight Center on November 3, 1999 [5].

6.6.2 Investigation

The failure of the composite tank is believed to be the result of multiple defect interactions in the sandwich structure [5]. Microcracking is believed to have occurred under the combined thermal and mechanical loads, allowing liquid, and gaseous hydrogen ingression from the inner facesheets. Since cracking did not occur to the same extent on the outer facesheets, increases in core pressure occurred due to conversion of the LH_2 to gaseous hydrogen. Secondly, the bond between the core and the carbon/epoxy facesheets was compromised due to poor bonding and core fuzz remnants. And finally, several pieces of foreign object debris (FOD) left during bonding are believed to have provided the initial disbond that precipitated the failure of the composite tank (Figure 6.7).

6.6.3 Lessons Learned

The investigation revealed several manufacturing flaws in the X-33 structure as the cause of the failure [6]. The case study shows the importance of having adequate quality control and process robustness that can preclude the presence of FOD in the composite structure. Fracture mechanics analyzes using finite element models of the debond configurations and the virtual crack closure technique were conducted to estimate the critical core pressure for unstable debond growth [5]. The fracture mechanics analyzes showed that coupled with high internal core pressure, a lower bondline toughness, a J-shaped FOD was the likely cause for the final failure of the X-33 composite LH_2 tank. Low bondline toughness can reduce the damage tolerance of the structure by not preventing the initiation and propagation of damage. Careful design is also necessary such that internal core pressures do not build up excessively in sandwich structures. Adequate venting is necessary such that pressure differentials does not result in the growth of delaminations.

Figure 6.7 Multi-lobed composite sandwich tank that failed causing NASA to ultimately cancel the X-33 program. Courtesy of NASA.

References

1 Public Utilities Commission of Ohio,10-0369-EL-BGN, C.N., Correspondence from EDP Renewables North America LLC Electronically Filed by Mr. Michael J. Settineri on Behalf of EDP Renewables North America LLC. 2012. The Public Utilities Commission of Ohio: Columbus, OH.

2 GL, D., Rotor blades for wind turbines, in DNVGL-ST-0376.2015, 2015. DNV-GL: Bærum, Akershus, Norway.

3 Seaton, C. and S. Richter, *DOT/FAA/TC-14/20* Nonconforming Composite Repairs: Case Study Analysis. 2014: Federal Aviation Administration, William J. Hughes Technical Center, Aviation Research Division Atlantic City International Airport New Jersey 08405.

4 Canada, T.S.B.o.., Aviation Investigation Report, Loss of Rudder in Flight Air Transat, Airbus A310–308 C-GPAT Miami, Florida. 2007: The Transportation Safety Board of Canada, Quebec, Canada.

5 Goetz, R., R. Ryan, and A. Whitaker, Final Report of the X-33 Liquid Hydrogen Tank Test Investigation Team. 2000. Marshall Space Flight Center, Huntsville, AL.

6 Niedermeyer, M. and P. Munafo, X-33 LH2 Tank Failure Investigation Findings, 2002. NASA Technical Reports, Document number 20010020398. NASA.

Glossary

α_1	Coefficient of thermal expansion in the lamina fiber direction
α_2	Coefficient of thermal expansion 90 degrees from the lamina fiber direction
A	Laminate extension-shear coupling matrix
AE	acoustic emission
Allowable	Material property values obtained from test that are typically A or B basis-based statistically derived quantities.
AML	Laminate configuration parameter
B	Laminate bending-extension coupling matrix
BVID	Barely visible impact damage.
Carbon fibers	Carbon fibers contain ~95% carbon and are carbonized at 982–1482°C (1800–2700°F).
Coupon	A small test specimen.
Creep	Time dependent strain occurring at constant stress.
Critical structure	A load bearing structure/element whose integrity is essential in maintaining flight loads or operations.
CTE or α	Coefficient of thermal expansion
CV	Coefficient of variation.
D	Laminate bending stiffness matrix
DCB	Double cantilever beam
Defect	An imperfection in the structure that can be quantitatively shown to cause failure.
Dents	Local depressions from processing, impact or local pressure.

Composite Structures: Effects of Defects, First Edition.
Rani Elhajjar, Peter Grant and Cindy Ashforth.
© 2019 John Wiley & Sons Ltd. Published 2019 by John Wiley & Sons Ltd.

Design value	Properties determined from test for material or structural details or elements. Often based on allowables and used in analysis to compute margins of safety.
Disbond	Any unbonded area occurring between a facesheet and honeycomb core or between doublers and core.
ϵ	Strain
ϵ_1	Normal strain in the fiber direction
ϵ_2	Normal strain transverse to the fiber direction
Element	A portion of a more complex structure.
E_x	Axial stiffness in the laminate x-direction
E_y	Axial stiffness in the laminate y-direction
E_{11}	Axial stiffness in the lamina fiber direction
E_{22}	Axial stiffness 90 degrees from the lamina fiber direction
Fracture toughness	A generic term used to refer to different fracture resistance measurements. The term refers to the structure's resistance to crack extension.
γ_{12}	Shear strain in the 1-2 direction
Galvanic corrosion	Accelerated corrosion in a metal due to contact with carbon-fiber composite.
Graphite fibers	Graphite fibers contain ~99% carbon and are first carbonized followed by graphitizing at temperatures between 1982° and 3037°C (3600° and 5500°F).
G_{xy}	Shear stiffness in the laminate x-y direction
G_{12}	Shear stiffness in the lamina 1-2 direction
Imperfection	A deviation from specifications that may or may not be a defect.
κ	Curvature
Knockdown factor	See retention factor.
K_t	Orthotropic elastic stress concentration factor near a hole
M	Moments
N	In-plane loads
ν_{12}	Poisson's ratio in the 1-2 direction
ν_{21}	Poisson's ration in the 2-1 direction
NDT/NDE/NDI	Nondestructive testing/Nondestructive evaluation/Nondestructive investigation. Terms are sometimes used interchangeably.

Out of autoclave composites	Composites produced without the aid of autoclaves.
PEEK	Polyetheretherketone
Poisson ratio	The absolute value of the ratio of the transverse strain to the axial strain obtained from a uniaxial test.
Porosity	A series of small discontinuities or voids. The outer periphery is usually outlined and the porosity is treated as a zone of reduced properties or an un unbonded area of similar size and shape.
PPS	Polyphenylenesulphide
Q	Lamina stiffness matrix
R-ratio	The ratio of the algebraic minimum fatigue load to the algebraic maximum fatigue load.
Retention factor	The retention or knockdown factor is the ratio that of the mechanical property accounting for the flaw to the mechanical property in the unflawed condition.
σ_1	Stress in the fiber direction
σ_2	Stress transverse to the fiber direction
Stress amplitude	Half the difference between the maximum and minimum stress in one cycle of a repeated stress.
Stress concentration factor	A multiplying factor for increased local stress associated with a defect or discontinuity such as a notch or a hole. It is typically expressed as the actual to the nominal stress at the location considered.
τ_{12}	Shear tress in the 1-2 direction
T_g	Glass transition temperature
Thermal spiking	Application of high temperatures for short durations above temperature operational limits.
Ultrasonic inspection	Nondestructive method in which beams of high-frequency sound waves are introduced into materials.
UT	Ultrasonic transmission
VID	Visible impact damage
Zero-volume disbond	A disbond between adhesive and adherend that has a zero volume. Also known as a "kissing bond."

Index

Composite Structures: Effects of Defects, First Edition.
Rani Elhajjar, Peter Grant and Cindy Ashforth.
© 2019 John Wiley & Sons Ltd. Published 2019 by John Wiley & Sons Ltd.